Enzyme Nanoparticles

Enzyme Nanoparticles

Preparation, Characterisation, Properties and Applications

Chandra S. Pundir
Emeritus Scientist-CSIR, Department of Biochemistry,
Maharishi Dayanand University,
Rohtak-124001 Haryana, India

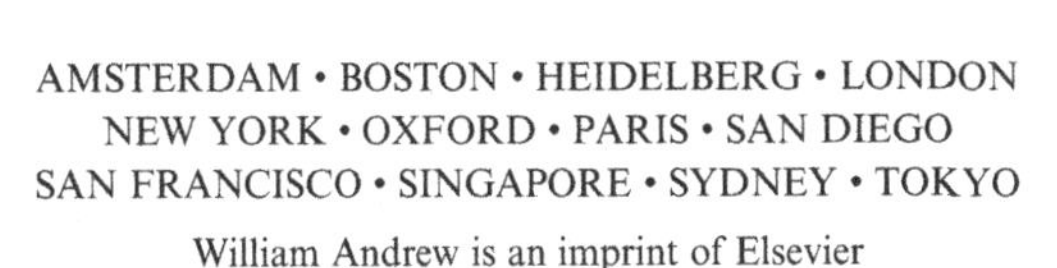

William Andrew is an imprint of Elsevier
225 Wyman Street, Waltham, MA 02451, USA
The Boulevard, Langford Lane, Kidlington, Oxford, OX5 1GB, UK

ISBN: 978-0-323-38913-6

Library of Congress Cataloging-in-Publication Data
A catalog record for this book is available from the Library of Congress

British Library Cataloguing-in-Publication Data
A catalogue record for this book is available from the British Library

For Information on all William Andrew publications
visit our website at http://store.elsevier.com/

DEDICATION

In the memories of my loving parents,

Late Mr. Prem Singh Pundir (Father)
and
Late Mrs. Rewa Devi (Mother)

DEDICATION

[illegible]

[illegible]

and

[illegible]

CONTENTS

ABOUT THE AUTHOR

Prof. Chandra S. Pundir has over 35 years (17 years as professor) of experience in teaching and research in Biochemistry/Biotechnology at PG level. He is the author of 221 research papers in national and international journals of repute (H-index: 20). Prof. Pundir obtained his master's and doctoral degree in Biochemistry in 1975 and 1980, respectively, from GB Pant University of Agriculture and Technology, Pantnagar, Uttrakhand, India. He has guided 207 Master Dissertations during 1987–2014 and entered into *Limca Book of Records* for guiding a maximum number of dissertations. He has also guided 28 PhD theses in Biochemistry/Biotechnology. He has completed 7 major Research projects and filed 6 Indian patents and 1 PCT patent as chief inventor. He is the member of several professional societies, including Royal Society of Chemistry, Oxford and Science peer reviewer of international journals including Research projects of Ministry of Business, Innovation and Employment (MBIE), New Zealand and Singapore Agencies for Science, Technology & Research (A*STAR). Presently, he is working as Emeritus Scientist of Council of Scientific & Industrial Research (CSIR), New Delhi, at the Department of Biochemistry, Maharishi Dayanand University (MDU), Rohtak, Haryana, India, after serving MDU for 30 years (1984–2014). His current research interest is in the area of Enzymology, Biosensor Technology, and Bionanotechnology.

PREFACE

Enzymes are specialized globular proteins (except ribozymes) that catalyse the chemical reactions occurring in living organisms, without being consumed in the reaction. These enzymes are synthesized in living cells only. After their isolation from the cells, the enzymes continue to function in vitro and thus used in various industrial processes. The diameter of these protein/enzymes is around 10 nm. The aggregated form of enzyme molecules in nanometer scale within the dimensions 10–100 nm is called enzyme nanoparticles (ENPs). These ENPs show their unique optical, electrical, electronic, thermal, chemical and catalytic (ability to facilitate electron transfer) besides increased surface area. The glutaraldehyde cross-linked and functionalized nanoparticles of few enzymes have been immobilized covalently, direct onto the surface of metal electrodes in the last 10 years, which has yielded improved biosensors. This new enzyme immobilization technology has opened a new area for biosensors, biomedical devices, diagnostics and enzyme therapy to improve their performance.

Among the more than 3000 enzymes known to exist in nature, the nanoparticles of very few (<10) enzymes have been prepared, characterized and employed in the construction of biosensors. Thus, ENP technology has a great scope for future research. However, no book is available in this rapidly emerging field. The present book entitled, "Enzyme nanoparticles: Preparation, characterization, properties and applications" is the first book in English on nanoparticles of enzymes with a focus on their preparation, characterization, immobilization and kinetic properties and applications of immobilized ENPs in the construction of improved biosensors for detection of various metabolites. Since a limited literature is available in this field, the book is based on few reports. It, therefore, will act as a concise information source for advanced studies in the field of ENP technology and provide invaluable source of coherent information for researchers, graduate and post graduate students in the fields of nanotechnology, biomedical engineering, bio-nanotechnology, pharmacy, dentistry, veterinary, life science and biochemistry.

This book would not have been possible without the help of ELSEVIER INC., Waltham, MA (the "Publisher"), its reviewers and editors, Drs. Jeffrey Freeland, Peter Gane and F. Hellwig and Simon Holt, Acquisitions Editor for Micro- and Nanotechnology books. I thank them for their encouragement, constructive suggestions and criticisms during the development of this book.

I acknowledge my wife Snehlata, my loving children Shikha, Saurabh, Ruchika and Abhishek and my adorable granddaughter Saamya for their moral support and patience during the preparation of this book.

Last, but not the least, I thank Ms. Vishakha Aggarwal, Senior Research Fellow in my research laboratory, for her technical help and valuable suggestions in the preparation of the manuscript.

Suggestions for improvement may be sent to pundircs@rediffmail.com

Prof. Chandra S. Pundir

CHAPTER 1

Introduction to Enzyme and Nanotechnology

WHAT ARE NANOPARTICLES AND NANOTECHNOLOGY?

To give an idea of how small things are, the average sizes of some entities are given in Table 1.1 which provides their relative sizes in the following order: Atom > Electron > DNA double helix > Protein/Enzyme > Cell > RBC. Materials when reduced down to 1–100 nm ($1-100 \times 10^{-9}$ m) in their dimension show drastic changes in respect of their physical, chemical, optical, magnetic, mechanical and electrical properties. All these properties lead to exciting applications of these nanomaterials in bioscience, medicine and environmental science, cosmetics, electronics, etc. that is called nanotechnology. In other words, nanotechnology is defined as the science and engineering that deals with the design, synthesis, characterization and applications of materials in devices and systems through control of matter on the nanometre scale

Enzyme Nanoparticles. DOI: http://dx.doi.org/10.1016/B978-0-323-38913-6.00001-5

Table 1.1 Average Size of Some Entities

Component	Approx. size (nm)
Atom	0.15
Electron	1
DNA (double helix)	2
Protein molecule	10
Cell	1000
RBC (width)	7000

(normally in the range of 1–100 nm). Different nanostructures have been investigated to determine their properties and possible applications. These structures include nanotubes, nanofibres, nanorods, nanoparticles (NPs) and thin films. Of these nanomaterials, NPs are the best studied. NPs are of two types: organic and inorganic. Among the organic NPs, proteins/enzymes NPs have attracted the attention of researchers during the past decade, specifically their application in the construction of improved biosensors/analytical devices.

WHAT ARE ENZYMES AND THEIR AGGREGATES?

Enzymes, also known as biological catalyst, are specialized globular proteins (with the exception of a few RNAs called ribozyme), which accelerate the rate of biochemical reaction up to 10^8 fold by lowering the activation energy (Ea) of the reaction without themselves being consumed in the reaction. Ea is the minimum amount of energy required to bring the substrate(s) from its ground state to transition state. The enzymes are required in small quantities for their biological functions and they possess characteristic kinetic parameters such as **optimum pH** (the most favourable pH at which the enzyme is most active), **incubation temperature** (the temperature at which the enzyme activity is maximum), **time of incubation** (time required to get the maximum activity of an enzyme), **Michaelis–Menten constant, Km** (the substrate concentration at half of its **maximal velocity (Vmax)**. The portion of enzyme where the substrate binds is called **active site**. Simple enzymes show hyperbolic relationship between their activity and substrate concentration and follow Michaelis–Menten kinetics during the reaction ($v = Vmax[S]/Km + [S]$; where v = initial velocity of reaction). The activity of enzymes can be increased or decreased by certain compounds known as activators and inhibitors, respectively. The inhibition of enzyme is of two types, irreversible and reversible. The reversible inhibition is classified into three types: (i) **competitive inhibition**, where both substrate and inhibitor (substrate analogue) compete for the same site, i.e. active site of the enzyme, (ii) **non-competitive inhibition**, where the inhibitor binds with the enzyme (E) or enzyme–substrate (E–S) complex or both at a site other than active site and (iii) **uncompetitive inhibition**, where the inhibitor binds only with the E–S complex. Enzymes are generally very specific, i.e. they act on a particular substrate. However, some enzymes have absolute specificity, while others have group-specificity, stereo-specificity and geometric specificity. Many enzymes require non-protein substances for their optimum activity, called **co-factors**. These co-factors are either inorganic or organic or both. The organic co-factors are named as **co-enzymes**, which are generally the derivatives of vitamins. The enzymes bind with their substrate (S) to form E–S complex, which is finally converted into product (P), due to favourable proximity and favourable orientation, acid–base catalysis, covalent catalysis, electrostatic catalysis, metal ion catalysis (in case of metalloenzymes), strain and distortion and the free enzyme is released. More than 3000 enzymes are known so far. Enzymes are classified into six major groups: **EC 1. Oxidoreductases**, which catalyse oxidation/reduction reactions, **EC 2. Transferases**, which catalyse the transfer of functional groups, **EC 3. Hydrolases**, which catalyse the hydrolysis of various bonds, **EC 4. Lyases**, which cleave various bonds, non-hydrolytically and nonoxidatively, **EC 5. Isomerases**, which catalyse isomerization within a single molecule, **EC 6. Ligases**, which join two molecules covalently at the expense of ATP/GTP. **Unit activity** (U) of an enzyme is defined as the amount of enzyme/protein required to convert one micro molecule of the S into P per min/ml. The **specific activity** is the ratio between activity and protein in milligrams. The activity of an enzyme is controlled by many factors such as restricted proteolysis, covalent catalysis (phosphorylation and de-phosphorylation, adenylation and de-adenylation, and acetylation and deacetylation), their end products, adenylate energy charge (ratio of ADP + AMP and ATP), induction and repression. The diameter of a protein/enzyme molecule is approximately 10 nm in solution. However, when the enzyme molecules are aggregated up to 100 nm, they exhibit unique properties such as high surface area, good stability, biocompatibility, but more importantly, high conductivity and sensitivity. These aggregates of enzyme molecules are called **enzyme nanoparticles (ENPs)**.

WHAT ARE ENZYME NANOPARTICLES?

ENPs can be defined as the clustered assembly of enzyme molecules in a fixed form of protein structure in nanometre dimension ranging between 10–100 nm. ENPs show their unique optical, electrical, electronic, thermal, chemical and mechanical, catalytic (ability to facilitate electron transfer) properties besides increased surface area. The ability to tailor the properties of nanoparticles/nanomaterials offers excellent properties for enhancing the performance of enzyme-based sensors. Direct attachment of proteins/ enzyme onto nanoparticles may cause their denaturation, thus loss of its activity. To overcome this problem, enzyme molecules have been aggregated to form their nanoparticles and cross-linked within their self in controlled manner, before their immobilization. This has resulted in promising strategy for development of biosensors with improved analytic performance not just in terms of detection limit but also a much larger current response (Sharma et al., 2011). Liu et al. (2005) prepared and characterized NPs of horse-radish peroxidase (HRP) for the first time. Thereafter, formation and application of ENPs of glucose oxidase (GOD), cholesterol oxidase (ChOx) and uricase in construction of amperometric biosensors were reported (Sharma et al., 2011; Kundu et al., 2012; Chawla et al., 2013; Chauhan et al., 2014).

The ENPs are different from **single enzyme nanoparticles (SENs)**, which are single enzyme molecules surrounded/ armoured by/with a nanoscale organic/inorganic macro-molecules network with enhanced catalytic activity and thermal stability. Kim and Grate (2003) prepared SENs of trypsin and chymotrypsin. The resulting SEN deeply

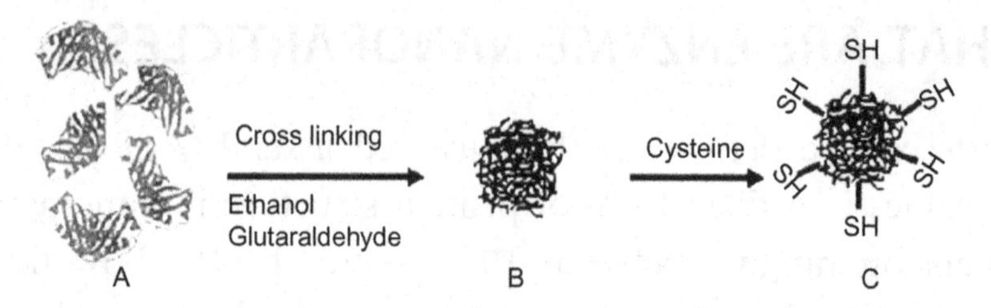

Figure 1.1 Scheme of enzyme nanoparticle preparation: A: free enzyme molecules; B: cross-linked enzyme nanoparticles; C: thiotinated enzyme nanoparticles. Source: Liu et al. (2005).

preserved the enzyme activity, while enhancing its thermal stability. SENs of HRP and cellulase were also reported recently to get the improved enzyme tolerance at elevated temperature and enhanced cellulose degradation, respectively (Khosravi et al., 2012; Blanchette et al., 2012).

HOW ARE NANOPARTICLES OF ENZYMES PREPARED?

The nanoparticles of the few enzymes such as HRP, GOD, ChOx and uricase have been prepared by desolvation and glutaraldehyde cross-linking method (Liu et al., 2005; Sharma et al., 2011; Kundu et al., 2012; Chawla et al., 2013; Chauhan et al., 2014) (Figure 1.1).

WHAT DOES THIS BOOK DESCRIBE?

This book describes in detail the preparation, characterization and immobilization of nanoparticles of HRP, GOD, ChOx and uricase and their kinetic properties, and applications of immobilized ENPs in fabrication of various bionanosensors. The future perspectives in ENP technology are also discussed briefly.

REFERENCES

Blanchette, C., Lacayo, C.I., Fischer, N.O., Hawang, M., Thelen, M.P., 2012. Enhanced cellulose degradation using cellulose nanospheres. PLOS One 7, e42116.

Chauhan, N., Kumar, A., Pundir, C.S., 2014. Construction of an uricase nanoparticles modified Au electrode for amperometric determination of uric acid. App. Biochem. Biotechnol. 174: 1683–1694.

Chawla, S., Rawal, R., Sonia, Ramrati, Pundir, C.S., 2013. Preparation of cholesterol oxidase nanoparticles and their application in amperometric determination of cholesterol. J. Nanopart. Res. 15, 1934–1943.

Khosravi, A., Vossoughi, M., Sharokhian, S., Alemzadeh, I., 2012. Synthesis and stability evaluation of HRP single enzyme nanoparticles. In: Proceedings of International Conference on Nanostructures (ICNS4), pp. 857–869.

Kim, J., Grate, J.W., 2003. Single enzyme nanoparticles armored by a nanometre scale organic/inorganic network. Nano Lett. 3, 1219–1222.

Kundu, N., Yadav, S., Pundir, C.S., 2012. Preparation and characterization of glucose oxidase nanoparticles and their application in DO metric determination of serum glucose. J. Nanosci. Nanotechnol. 13, 1710–1716.

Liu, G., Lin, Y., Ostatna, V., Wang, J., 2005. Enzyme nanoparticles based electronic biosensor. Chem. Commun. 27, 3481–3483.

Sharma, S., Shrivastav, A., Gupta, N., Srivastava, S., 2011. Amperometric biosensor: increased sensitivity using enzyme nanoparticles. In: 2010 International Conference on Nanotechnology and Biosensors, IPCBEE, vol. 2, pp. 24–26.

CHAPTER 2

Preparation of Enzyme Nanoparticles

METHOD OF ENPs FORMATION

Nanoparticles (NPs) of soluble proteins were prepared by aggregating proteins (bovine serum albumin, BSA) into NPs by (1) emulsification in plant oil (Scheffel et al., 1972), (2) desolvation by ethanol or natural salts followed by glutaraldehyde cross-linking (Coester et al., 2000), (3) simple coacervation by anhydrous ethanol, glutaraldehyde and ethanolamine (Sailaja et al., 2011) and (4) cross-linking in water and oil emulsion under high pressure (Ezpeleta et al., 1999). However, the enzyme nanoparticles (ENPs) have been prepared by **desolvation** with ethanol and subsequent cross-linking with glutaraldehyde so far (Liu et al., 2005). The details of other methods are described at the end of this chapter. The **desolvation method** includes the following steps:

1. **Desolvation:** Dissolve solid enzyme in reaction buffer at a concentration of 1–2 mg/ml. Add desolvating agent,

Enzyme Nanoparticles. DOI: http://dx.doi.org/10.1016/B978-0-323-38913-6.00002-7

i.e. absolute ethanol into dissolved enzyme in 1:2 ratio, dropwise at a rate of 0.1–0.5 ml per minute under continuous stirring at a speed of 500 rpm or 800 *g* in cold (4°C). The desolvating agent encourages the enzyme/protein molecules to aggregate into small particles or NPs (size: 1–100 nm) by removing water molecules in between protein molecules/ions, thus reducing the distance between enzyme molecules leading to the formation of their clusters.

2. **Cross-linking:** After the desolvation process, add 2.5% to 8.0% glutaraldehyde (1.8 ml) into ENPs suspension in buffer (9.0 ml) under continuous stirring at 500 rpm at 4°C in an ice bath for 24 hours, ensuring the cross-linking of ENPs and thus forming NPs of enzyme. Such a high concentration of glutaraldehyde is likely to provide intermolecular cross-linking of ENPs through Schiff base.
3. **Purification:** Separate the resulting cross-linked ENPs from free enzyme molecules by centrifuging the suspension at 14,000 *g* for 10 minutes at 4°C and redisperse the pellet to original volume in buffer by ultrasonication for 5 minutes. Repeat the process two more times. Finally, separate the ENPs from enzyme solution by centrifugation at 12,000 rpm/15,000 *g* for 10 minutes at 4°C and disperse the NPs in buffer and sonicate the pellet in buffer for 5 minutes and store it at 4°C.
4. **Functionalization:** Add solid cysteamine dihydrochloride (0.12 gm) or cysteine (0.12 gm) into ENPs suspension (10.8 ml) to introduce $-NH_2$ groups or $-SH$ groups on the surface of ENPs, under continuous stirring.

CHEMISTRY OF FORMATION AND FUNCTIONALIZATION OF ENPs

The aggregates of enzyme molecules are formed under the dropwise addition of ethanol at 4°C due to decrease of hydration layer around enzyme molecules. The enzyme molecules interact with each other by forces like van der Waals, hydrophobic and electrostatic interaction resulting in the formation of ENPs aggregates.

1. **Cross-linking of ENPs:** The two aldehyde groups on both the ends of glutaraldehyde get attached to free $-NH_2$ group of lysine residue on the surface of ENPs through Schiff base formation as given below:

$$\text{ENP-NH}_2 + \text{OHC-(CH}_2)_3\text{-CHO} + \text{H}_2\text{N-ENPs} \longrightarrow$$
$$\underbrace{\text{ENP-N}{=}\text{CH-(CH}_2)_3\text{-CH}{=}\text{N-ENP}}_{\text{Schiff base}} + 2\text{H}_2\text{O}$$

2. **Functionalization of ENPs:** Addition of cysteine into ENPs suspension introduces $-SH$ groups, while cysteamine introduces $-NH_2$ groups on the surface of cross-linked ENPs as shown in Figures 2.1 and 2.2, respectively.

The desolvation method has been employed for the preparation of NPs of few enzymes such as horseradish peroxidase (Liu et al., 2005), glucose oxidase from *Aspergillus niger* (Sharma et al., 2011; Kundu et al., 2012), cholesterol oxidase from microorganism sp. (Chawla et al., 2013) and uricase from *Candida* sp. expressed in *Escherichia coli* (Chauhan et al., 2014). The other

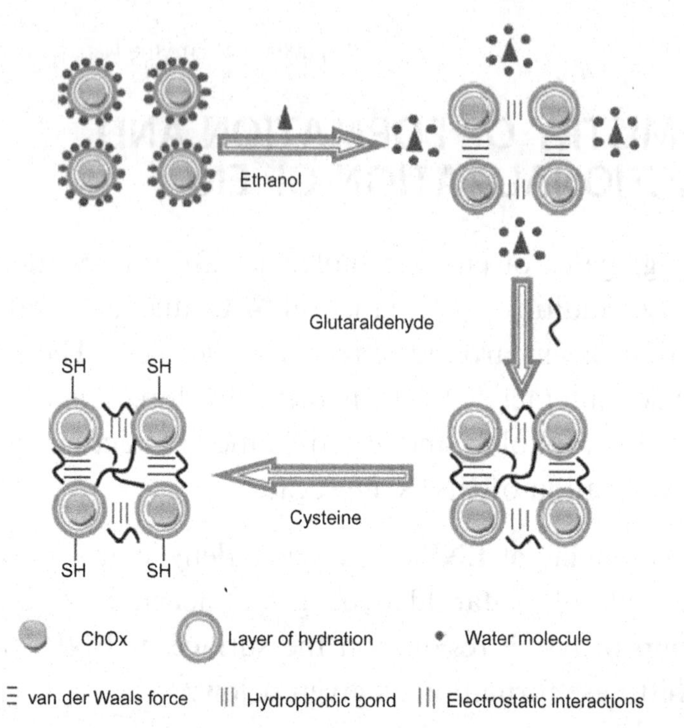

Figure 2.1 Schematic diagram of formation of cholesterol oxidase nanoparticles (ChOx NPs) by desolvation method and introduction of −SH groups by cysteine on the surface of ChOx NPs. Source: Chawla et al. (2013).

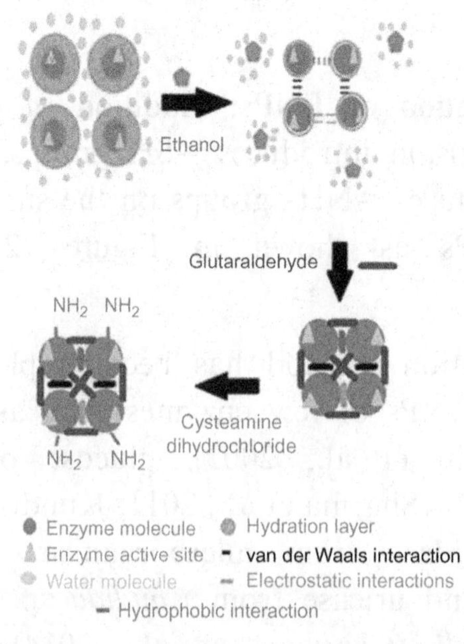

Figure 2.2 Schematic diagram of formation of glucose oxidase nanoparticles (GOD NPs) by desolvation method and introduction of $-NH_2$ groups by cysteamine dihydrochloride on the surface of GOD NPs. Source: Kundu et al. (2012).

methods for formation of NPs of soluble proteins are discussed below:

1. **Emulsification method:** In this process, an aqueous solution of albumin is turned into an emulsion at room temperature in plant oil (cottonseed oil) by a mechanical homogenizer with high speed to obtain a homogeneous emulsion. This method would create high dispersion within the particles. The above emulsion is added to a high volume of preheated oil (over 120°C) drop by drop. This process results in rapid evaporation of existing water and irreversible aggregation of albumin molecules. This process generates NPs (Scheffel et al., 1972).
2. **Simple coacervation method:** It is implemented for preparation of BSA NPs (Sailaja et al., 2011). Anhydrous ethyl alcohol is added to 150 ml BSA (5 mg/l in 10 mM Tris/HCl containing 0.02% sodium azide, pH 7.5) till the solution became turbid, then 150 μl of 25% glutaraldehyde is added for cross-linking. The reaction is allowed to continue at room temperature (24°C). Ethanolamine is added to block the non-reacted aldehyde functional group. Tween-20 is then added at a final concentration of 0.01% (v/v) to stabilize the preparation. Large aggregates/NPs are eliminated by centrifugation (50,000 *g*, 30 min, 4°C). The supernatant is dialysed and subsequently micro and ultrafiltered through a 0.2 μm acetate membrane and polyvinyl chloride copolymer membrane with a cut-off of 300 kDa, respectively.
3. **Cross-linking in w/o emulsion:** The method involves the emulsification of BSA/human serum albumin (HSA) or

protein aqueous solution in oil using high pressure homogenization. The water-in-oil (w/o) emulsion so formed is then poured into preheated oil. The suspension in preheated oil maintained above 100°C is held stirred for a specific time in order to denaturate and aggregate the protein contents of aqueous pool completely and to evaporate water. Proteinaceous subnanoscopic particles are thus formed, where the size of the internal phase globule mainly determines the ultimate size of particulates (Ezpeleta et al., 1999).

SUMMARY

Nanoparticles of enzymes are prepared by functionalization using ethanol, cross-linked by glutaraldehyde and functionalized by introducing $-SH/-NH_2$ groups with the treatment of cysteine/cysteamine.

REFERENCES

Chauhan, N., Kumar, A., Pundir, C.S., 2014. Construction of an uricase nanoparticles modified Au electrode for amperometric determination of uric acid. App. Biochem. Biotechnol. 174: 1683–1694.

Chawla, S., Rawal, R., Sonia, Ramrati, Pundir, C.S., 2013. Preparation of cholesterol oxidase nanoparticles and their application in amperometric determination of cholesterol. J. Nanopart. Res. 15, 1934–1943.

Coester, C.J., Langer, K., Von, B., 2000. Gelatin nanoparticles by two step desolvation: a new preparation method, surface modification and cell uptake. J. Microencapsul. 2, 187–193.

Ezpeleta, I., Irchae, J.M., Stainmesses, S., Chabenat, C., Popineau, Y., Orecchionic, A.M., 1999. Preparation of *Ulex* europaeus lectin-gliadin nanoparticles conjugates and their interaction with gastrointestinal mucus. Int. J. Pharm. 191, 25–32.

Kundu, N., Yadav, S., Pundir, C.S., 2012. Preparation and characterization of glucose oxidase nanoparticles and their application in DO metric determination of serum glucose. J. Nanosci. Nanotechnol. 13, 1710–1716.

Liu, G., Lin, Y., Ostatna, V., Wang, J., 2005. Enzyme nanoparticles based electronic biosensor. Chem. Commun. 27, 3481–3483.

Sailaja, A.K., Amareshwar, P., Chakravarty, P., 2011. Different techniques used for the preparation of nanoparticles using natural polymers and their application. Int. J. Pharm. Pharm. Sci. 3, 45–50.

Scheffel, U., Rhodes, B.A., Natrajan, T.K., Wagner, H.N., 1972. Albumin microspheres for study of reticuloendothelial system. J. Nucl. Med. 13, 498–503.

Sharma, S., Shrivastav, A., Gupta, N., Srivastava, S., 2011. Amperometric biosensor: increased sensitivity using enzyme nanoparticles. In: 2010 International Conference on Nanotechnology and Biosensors, IPCBEE, vol. 2, pp. 24–26.

CHAPTER 3

Characterization of Enzyme Nanoparticles

Ultraviolet (UV) spectroscopy, transmission electron microscopy (TEM) and Fourier transform infrared (FTIR) spectroscopy are employed to characterize the following properties of enzyme nanoparticles (ENPs) aggregates:

1. **Shape and size:** The TEM images of aggregates of ENPs have revealed their spherical shape and size with a diameter in the range of 100–200 nm (Figure 3.1). However, the diameter of free/monomeric enzyme molecules is in the range of 2–18 nm (Table 3.1).
2. **Colour:** Enzymes may or may not acquire the same colour after their aggregation into nanoparticles. ENPs aggregates of horseradish peroxidase (HRP) are seemingly white in colour, resulting from transparency of protein to the electron beam (Figure 3.1). The ENP aggregates could also acquire their colour as that of its free/ionized form.

Enzyme Nanoparticles. DOI: http://dx.doi.org/10.1016/B978-0-323-38913-6.00003-9

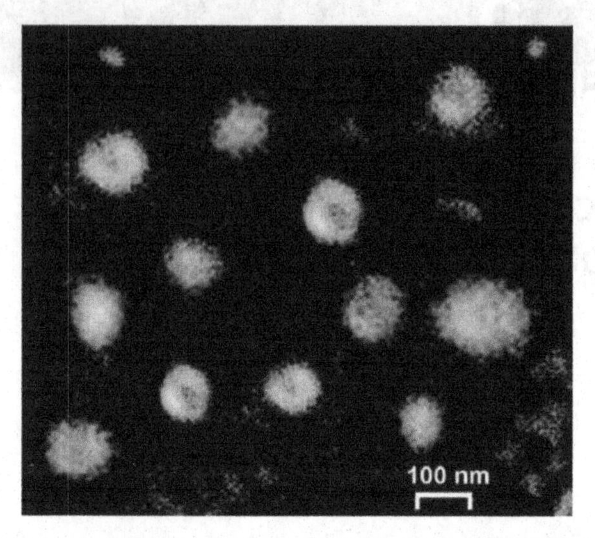

Figure 3.1 Transmission electron microscopic (TEM) images of horseradish peroxidase nanoparticles depicting their shape, size and colour. Source: Liu et al. (2005).

Table 3.1 Size of Enzyme Nanoparticles (ENPs) and Free Enzymes (in nm)

ENPs	Diameter of ENPs	Diameter of Free Enzyme	Reference
Horseradish peroxidase	100	14.2	Liu et al. (2005)
Glucose oxidase	117	2–8	Kundu et al. (2012)
Cholesterol oxidase	100–200	–	Chawla et al. (2013)
Uricase	100	14–18	Chauhan et al. (2014)

3. **UV absorption spectra:** Free enzyme, e.g. glucose oxidase (GOD), shows the characteristic well defined into two absorption peaks: peak I at 235 nm, due to absorption by peptide bond; and peak II at 320 nm, due to aromatic amino acids of protein/free enzyme in buffer (0.1 M phosphate buffer, pH 7.4). These peaks increase after the

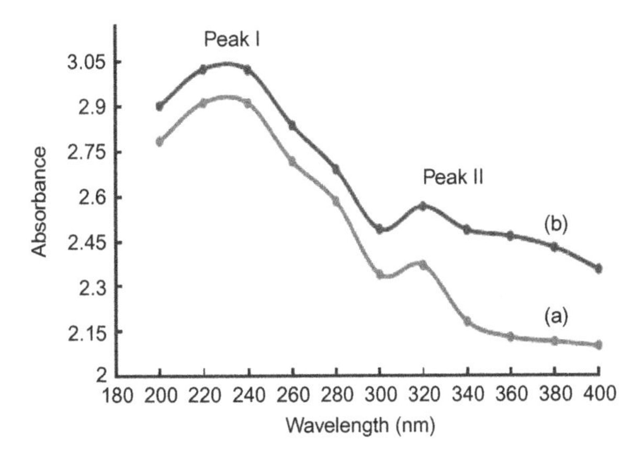

Figure 3.2 UV spectra of free glucose oxidase (GOD) (curve a) and GOD nanoparticles aggregates (curve b). Source: Chawla et al. (2013).

formation of NPs (Figure 3.2). However, GOD NPs aggregates show absorption maxima at 235 nm compared to that of free GOD at 267 nm (Figure 3.3a and b). This shift, known as Blue shift, is due to the formation of NPs (Sharma et al., 2011).

4. **FTIR spectra:** An FTIR spectral study of GOD NPs aggregates disclose a vibration band at 1673.20 cm^{-1} due to stretching of C = N bond after glutaraldehyde cross-linking along with vibration band of amide I band (1700 – 1600 cm^{-1}) caused by C=O stretching vibration of peptide linkages and amide II band (1600 – 1500 cm^{-1}) in enzyme backbone resulting from combination of N–H bond in plane bending and C–N stretching of peptide group. This spectra of functionalized GOD NPs aggregates also revealed broadening of vibration band at 3000–3500 cm^{-1} as compared to free enzyme molecules confirming more free $-NH_2$ groups introduced by cysteamine dihydrochloride on GOD NPs aggregates (Kundu et al., 2012) (Figure 3.4).

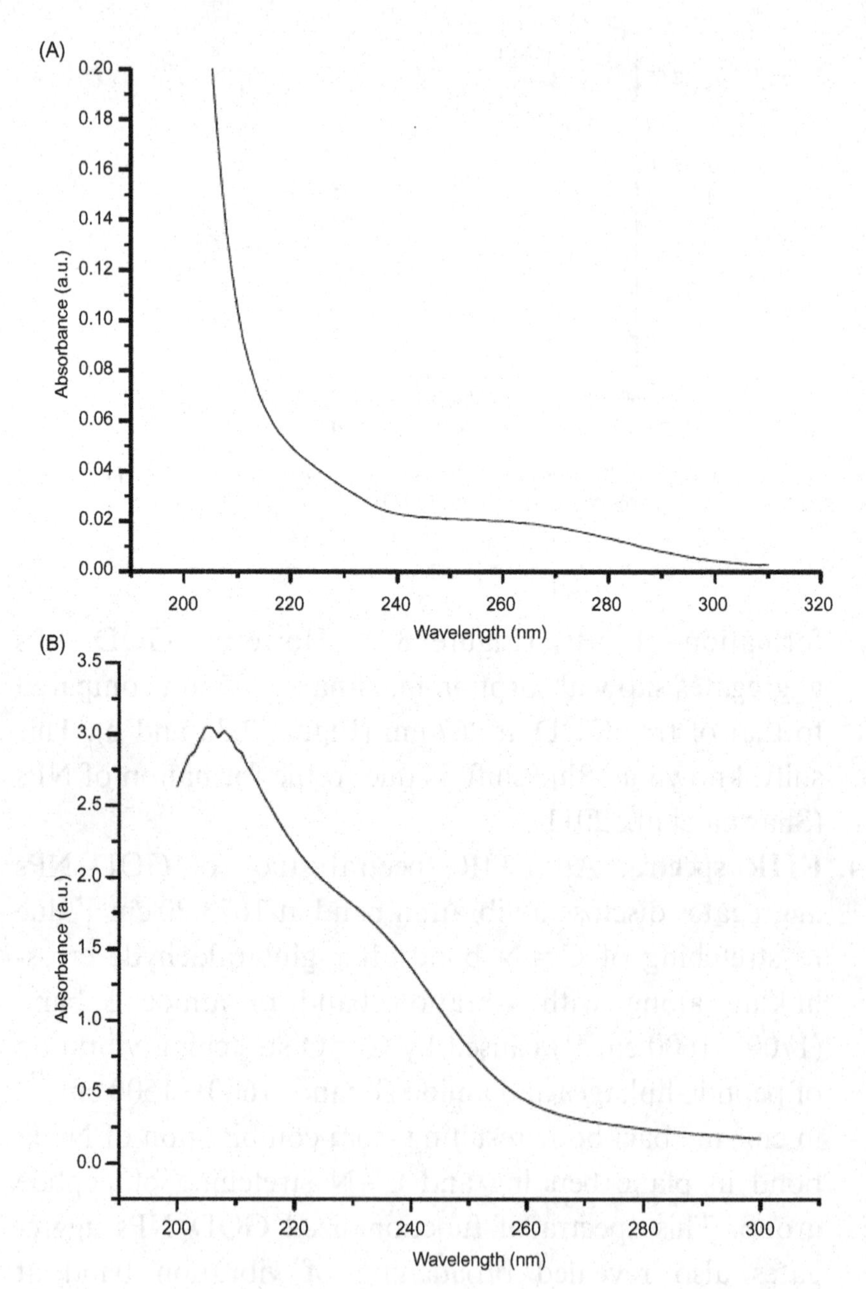

Figure 3.3 (A) UV visible spectra of free glucose oxidase; (B) UV visible spectra of free glucose oxidase nanoparticles. Source: (A) and (B) Sharma et al. (2011).

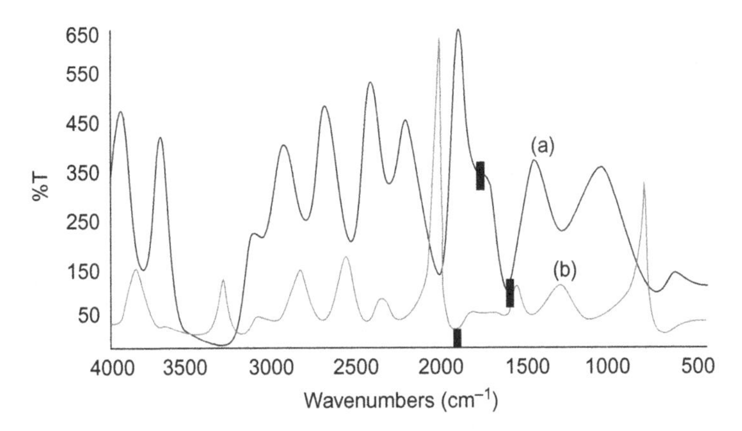

Figure 3.4 FTIR spectra of free glucose oxidase (GOD) (curve a) and GOD nanoparticles aggregates (curve b). Source: Kundu et al. (2012).

SUMMARY

ENPs are characterized by TEM study, UV spectroscopy and FTIR spectroscopy which revealed their spherical shape with diameter in the range of 100–200 nm, increased absorption in UV range and C=N stretching due to their glutaraldehyde cross-linking.

REFERENCES

Chauhan, N., Kumar, A., Pundir, C.S., 2014. Construction of an uricase nanoparticles modified Au electrode for amperometric determination of uric acid. App. Biochem. Biotechnol. 174, 1683–1694.

Chawla, S., Rawal, R., Sonia, Ramrati, Pundir, C.S., 2013. Preparation of cholesterol oxidase nanoparticles and their application in amperometric determination of cholesterol. J. Nanopart. Res. 15, 1934–1943.

Kundu, N., Yadav, S., Pundir, C.S., 2012. Preparation and characterization of glucose oxidase nanoparticles and their application in DO metric determination of serum glucose. J. Nanosci. Nanotechnol. 13, 1710–1716.

Liu, G., Lin, Y., Ostatna, V., Wang, J., 2005. Enzyme nanoparticles based electronic biosensor. Chem. Commun. 27, 3481–3483.

Sharma, S., Shrivastav, A., Gupta, N., Srivastava, S., 2011. Amperometric biosensor: increased sensitivity using enzyme nanoparticles. In: 2010 International Conference on Nanotechnology and Biosensors, IPCBEE. vol. 2, pp. 24–26.

CHAPTER 4

Immobilization of Enzyme Nanoparticles

Immobilization of enzyme can be defined as confinement/attachment of the enzyme in/onto an organic/inorganic support, physically or chemically or both in such a manner that it retains its activity and could be reused. Enzymes have been immobilized onto different supports by adsorption, entrapment, microencapsulation, covalent coupling, metal binding and cross-linking through bi-functional reagents. Enzyme nanoparticles (ENPs) aggregates are immobilized covalently onto both organic and inorganic supports after their functionalization to get the greater productivity, as the ENPs can be reused over a long period, with more precise control of the extent of reaction, capability of automation and continuous operation and elimination of enzyme inactivation. This chapter presents the process of covalent immobilization of ENPs aggregates of horseradish peroxidase (HRP) onto Au electrode, glucose oxidase (GOD) from *Aspergillus niger* onto Pt electrode and nitrocellulose

Enzyme Nanoparticles. DOI: http://dx.doi.org/10.1016/B978-0-323-38913-6.00004-0

(NC) membrane, cholesterol oxidase (ChOx) from microorganisms and uricase from *Candida* onto Au electrode.

IMMOBILIZATION OF HRP NPs AGGREGATES

Liu et al. (2005) reported for the first time the covalent immobilization of thiolated HRP NPs aggregates onto Au electrode, as follows:

1. Scan the potential of the cleaned bare Au electrode over 0.5–1.5 V range in freshly prepared 0.2 M H_2SO_4 until the voltammogram characteristic of the clean polycrystalline Au is established.
2. Place the polycrystalline gold electrode in the HRP NPs suspension under mild stirring at 4°C for 12 h to give a nanoparticle self-assembled HRP NPs layer.
3. Rinse the HRP NPs bound Au electrode (HRP NPs/AuE) with a 0.1 M phosphate buffer (PB, pH 7.4) carefully and store in a PB buffer at 4°C, when not in use.

IMMOBILIZATION OF GOD NPs AGGREGATES

GOD NPs aggregates were immobilized covalently onto Pt electrode and NC membrane by Sharma et al. (2011) and Kundu et al. (2012) as given below.

Onto Pt Electrode

1. Clean Pt electrode by dipping it into boiling H_2SO_4 for 30 min followed by washing with distilled water (DW).

2. Amino functionalize the cleaned electrode by dipping it into aqueous cysteamine dihydrochloride solution (10 mg/ml) in dark for 2 h.
3. Immerse the electrode into 10% (v/v) glutaraldehyde for 20 min. Wash it thoroughly with DW.
4. Immobilize GOD NPs onto the glutaraldehyde activated Pt electrode by immersing it into the ENPs suspension for 10 h at 4°C.
5. Rinse this GOD NPs bound Au electrode with DW. Dry it in air and store in 0.1 M phosphate buffer (pH 7.4) at 4°C.

Onto NC Membrane

1. Wash NC membrane ($1 \times 1.5\ cm^2$) with DW to remove any debris and then dry it in air for 30 min.
2. Treat the membrane with 0.5% chitosan (in 2% acetic acid) for 24 h at room temperature to introduce $-NH_2$ groups on its surface.
3. Remove the unbound $-NH_2$ groups by washing with 10% methanol followed by air drying for 30 min.
4. Activate the membrane by immersing it in 2.5% glutaraldehyde (in 0.1 M PB, pH 6.0) for 2 h at room temperature.
5. Wash the membrane thoroughly with 0.1 M PB, pH 6.0.
6. Spread GOD NPs aggregates suspension (0.5 ml) onto activated NC membrane uniformly and keep it overnight at 4°C for their covalent immobilization.
7. Wash the GOD NPs aggregates bound NC membrane with 0.1 M PB, pH 6.0.

8. Discard the GOD NPs aggregates and test both NC membranes and discard for enzyme activity.
9. Mount GOD NPs bound NC membrane over the sensing part of combined electrode of dissolved oxygen (DO) meter with a parafilm and connect it to the main apparatus of DO meter.

IMMOBILIZATION OF ChOx NPs AGGREGATES

Thiolated ChOx NPs aggregates were immobilized covalently onto Au electrode by Chawla et al. (2013) as given below:

1. Clean Au wire with piranha solution (H_2SO_4:H_2O_2 in 3:1 ratio) for 20 min, then rinse thoroughly with DW.
2. Polish the Au electrode with alumina slurry.
3. Layer/mount 2 ml suspension of thiolated ChOx NPs in 0.1 M potassium PB, pH 7.0, onto surface of polished Au electrode and keep it overnight at 4°C for immobilization.
4. Wash the resulting enzyme electrode with immobilized ChOx NPs 3–4 times with same buffer solution, to remove residual unbound protein.
5. Store the ChOx NPs bound Au electrode/working electrode at 4°C, when not in use.

IMMOBILIZATION OF URICASE NPs AGGREGATES

Thiolated uricase NPs aggregates were immobilized covalently onto Au electrode by Chauhan et al. (2014) as follows:

1. Clean bare Au electrode as given above and scan it over the range 0.5–1.5 V in freshly prepared 0.2 M H_2SO_4 until the voltammogram characteristic of the cleaned polycrystalline Au is established.
2. Place the polycrystalline Au electrode in the uricase NPs suspension under mild stirring at 4°C for 12 h to provide a nanoparticles self-assembled uricase NPs layer.
3. Rinse the uricase NPs bound Au electrode/working electrode with 50 mM of PB, pH 7.4, carefully and store it in the same buffer at 4°C until use.

CHEMISTRY OF IMMOBILIZATION OF ENPs

On Au Electrode

Thiol groups on the surface of ENPs provide a facile method for attaching the ENPs on the surface of Au electrode to prepare an ENPs-bound electrode and thus circumventing complications associated with solution system. The scanning of Au electrode in the range of 0.5–1.5 V in freshly prepared 0.2 M H_2SO_4 provides polycrystalline Au electrode. The thiol-functionalized cross-linked ENPs get bound to polycrystalline Au electrode

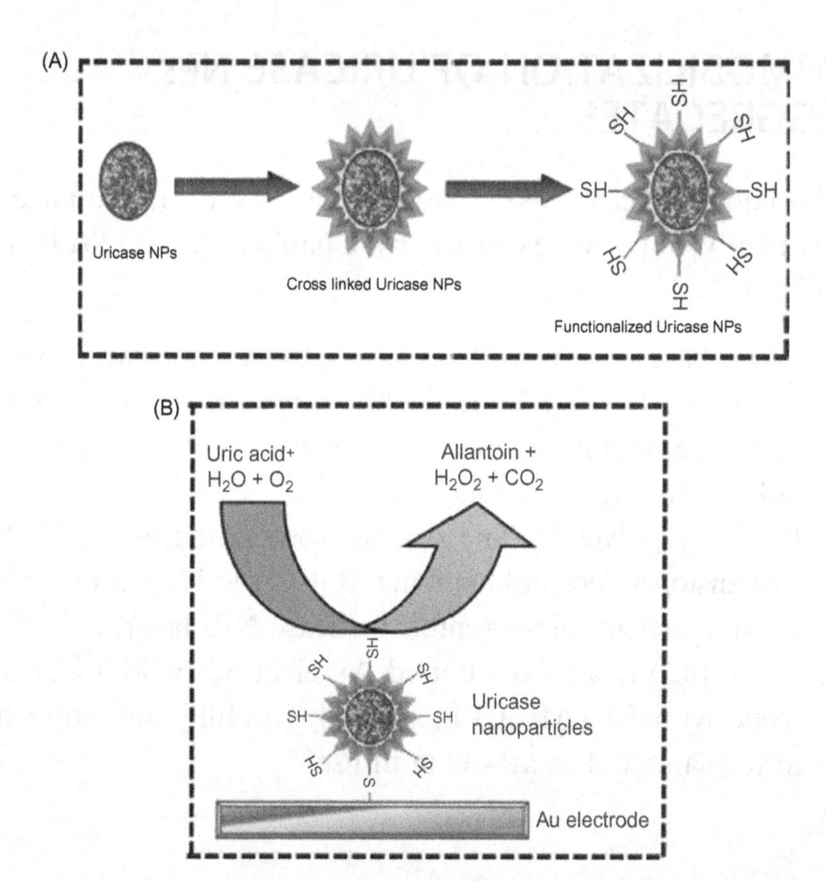

Figure 4.1 (A) Functionalization of uricase nanoparticles (NPs) and (B) immobilization of uricase NPs on thiolated Au electrode along with its reaction. Source: Chauhan et al., 2014.

through Au-thiolated bond using the reduction process as reported for uricase nanoparticles (UOxNPs) (Chauhan et al., 2014; Figure 4.1)

$$\underset{\text{Polycrystalline-AuE}}{\text{Au}_n\text{-AuO}} + \underset{\text{Thiolated UOxNPs}}{\text{2(UOxNPs)-SH}} \xrightarrow{\text{Reduction}} \underset{\text{UOx-NPs bound-AuE}}{\text{Au}_n\text{(Au-S-UOxNPs)}_2} + H_2O$$

On NC Membrane/Pt Electrode

The treatment of NC membrane with chitosan or Pt electrode with cysteamine HCl introduces $-NH_2$ groups on their surface. ENPs aggregates are immobilized covalently onto amino functionalized NC membrane/Pt electrode through glutaraldehyde coupling. Firstly, the NC membrane/Pt electrode is activated by reacting its free $-NH_2$ groups with $-CHO$ group of bi-functional glutaraldehyde at its one end. Now the free $-CHO$ groups of glutaraldehyde on its other end reacts with $-NH_2$ groups on the surface of ENPs aggregates forming the Schiff base between glutaraldehyde-activated NC membrane/Pt electrode and ENPs aggregates (Figure 4.2).

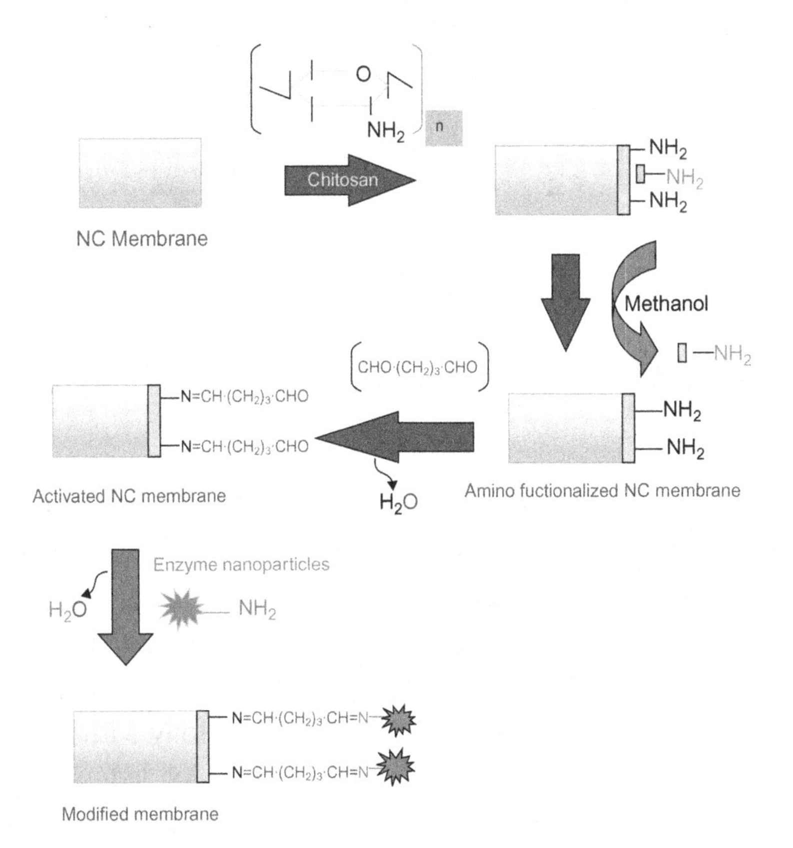

Figure 4.2 Scheme of chemical reactions of immobilization of glucose oxidase nanoparticles (GOD NPs) aggregates on nitrocellulose (NC) membrane. Source: Kundu et al., 2012.

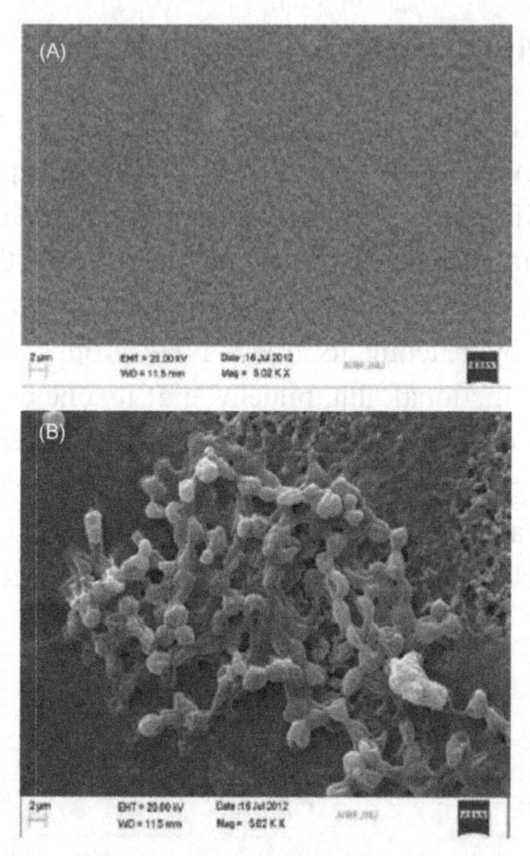

Figure 4.3 SEM images of NC membrane without (A) and with immobilized glucose oxidase aggregates (B). Source: Kundu et al., 2012.

CONFIRMATION OF IMMOBILIZATION OF ENPs

The immobilization of ENPs is confirmed by high resolution scanning electron micrographs (SEM) of support surface without and with the immobilized ENPs aggregates. The SEM image of bare support (metal electrode/membrane) shows the homogeneous granular surface, while the ENPs bound/modified support show globular structural morphology with clusters of aggregation of ENPs along with some beaded structure throughout the surface, confirming successful immobilization of ENPs aggregate (Figures 4.3 and 4.4).

Figure 4.4 SEM images of Au electrode without (A) and with immobilized cholesterol oxidase aggregates (B). Source: Chawla et al., 2013.

CONJUGATION YIELD AND % RETENTION OF IMMOBILIZED ENPs

A conjugation yield of 5.9 $\mu g/cm^2$ with 55.15% retention of initial activity of GOD NPs on NC membrane are reported (Kundu et al., 2012).

SUMMARY

ENPs aggregates are immobilized covalently onto thiolated Au electrode through Au-thiol bond and amino functionalized NC membrane/Pt electrode through glutaraldehyde coupling.

REFERENCES

Chauhan, N., Kumar, A., Pundir, C.S., 2014. Construction of an uricase nanoparticles modified Au electrode for amperometric determination of uric acid. App. Biochem. Biotechnol. http://dx.doi.org/10.1007/s12010-014-1097-6.

Chawla, S., Rawal, R., Sonia, Ramrati, Pundir, C.S., 2013. Preparation of cholesterol oxidase nanoparticles and their application in amperometric determination of cholesterol. J. Nanopart. Res. 15, 1934–1943.

Kundu, N., Yadav, S., Pundir, C.S., 2012. Preparation and characterization of glucose oxidase nanoparticles and their application in DO metric determination of serum glucose. J. Nanosci. Nanotechnol. 13, 1710–1716.

Liu, G., Lin, Y., Ostatna, V., Wang, J., 2005. Enzyme nanoparticles based electronic biosensor. Chem. Commun. 27, 3481–3483.

Sharma, S., Shrivastav, A., Gupta, N., Srivastava S., 2011. Amperometric biosensor: increased sensitivity using enzyme nanoparticles. In: 2010 International Conference on Nanotechnology and Biosensors, IPCBEE, vol. 2, pp. 24–26.

CHAPTER 5

Kinetic Properties of Enzyme Nanoparticles

Generally, nanoparticles of enzymes show their enhanced activity, because of higher surface area, better stability and biocompatibility. Although the size, shape, colour and chemical bonding of enzyme nanoparticles (ENPs) aggregates have been studied, their kinetic properties in native form have not been reported. However, the kinetic properties of ENPs of a few enzymes have been studied after their covalent immobilization onto artificial membrane (nitro-cellulose membrane) or metal wire (Au/Pt). A comparison of kinetic properties of immobilized ENPs with those of free enzymes (Table 5.1) have revealed the following changes in the properties of free enzyme after the preparation of their nanoparticles and subsequent immobilization.

Enzyme Nanoparticles. DOI: http://dx.doi.org/10.1016/B978-0-323-38913-6.00005-2

Table 5.1 A Comparison of Kinetic Properties of Immobilized Enzyme Nanoparticles (NPs) with Those of Free Enzymes

	Glucose Oxidase from *A. niger*		**Cholesterol Oxidase from Microorganism**		**Uricase from *Candida* sp.**	
Property	**Free**	**Immobilized NPs**	**Free**	**Immobilized NPs**	**Free**	**Immobilized NPs**
Support used	–	NC membrane	–	Au wire	–	Au wire
Method of immobilization	–	Covalent	–	Covalent	–	Covalent
Type of electrode	–	DO metric	–	Amperometric	–	Amperometric
Optimum pH	5.5	6.0	5.5	6.0	8.0	8.5
Optimum temperature (°C)	35	35–45	37	35	30	40
Incubation time	120 s	60 s	300 s	8 s	300 s	7 s
Km (mM)	6.2	4.17	–	ND	0.056	0.058
Linearity (mM)	Up to 13.8	0.056–1.39	Up to 6.4	0.69–38.88	Up to 1.5	0.005–0.8

Limit of detection (mM)	0.138*	0.55	0.01*	0.04	160	0.005
Half life at 4°C in days	180	180	60	90	60	210
No. of reuses	–	150	–	180	–	200
Reference	Kalia et al. (1999)	Kundu et al. (2012)	Varma and Nene (2003)	Chawla et al. (2013)	Freitas et al. (2009)	Chauhan et al. (2014)
	*Cayman chemical enzo kit for glucose		*Cell bio lab inc. Enzo kit for cholesterol		*Biovision Enzo kit for uric acid	

ND = not detected

OPTIMUM pH

Immobilized ENPs show slight increase in their optimum pH compared to that of free enzyme, for example, an increase of pH 0.4 for horseradish peroxidase (Liu et al., 2005), pH 0.5 for glucose oxidase (GOD) (Figure 5.1; Kundu et al., 2012), cholesterol oxidase (ChOx) (Chawla et al., 2013) and uricase (Chauhan et al., 2014). This change in optimum pH might be due to altered confirmation of ENPs after their immobilization.

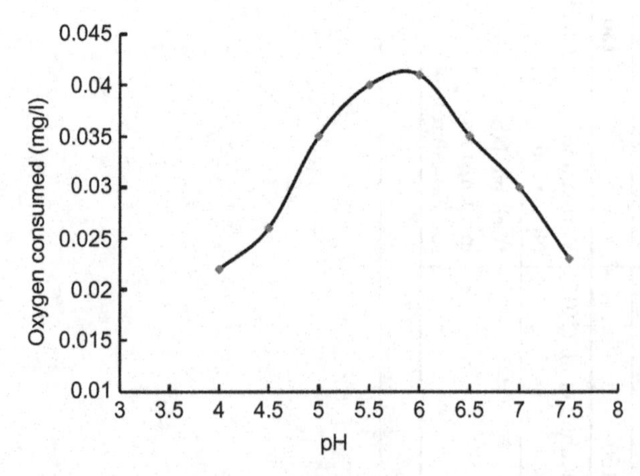

Figure 5.1 Effect of pH of reaction buffer on activity of glucose oxidase nanoparticles (GOD-NPs) immobilized onto nitro-cellulose (NC) membrane. The enzyme electrode/ O_2 electrode mounted with NC membrane bound GOD was immersed into 4.75 ml of 0.05 M PB, pH varying from 4.0–7.5, saturated with O_2 bubbling in a 10 ml glass beaker and dissolved O_2 was measured by DO meter. The meter reading was kept on hold and then 0.25 ml glucose solution (360 mg/dl) was added, shaken gently and dissolved O_2 consumed was measured again when reading becomes constant. Difference between two readings provided amount of dissolved O_2 consumed in enzyme assay. Source: Kundu et al., 2012.

OPTIMUM TEMPERATURE AND THERMOSTABILITY

The optimum temperature of immobilized ENPs exhibit a substantial increase compared to that of free enzyme, for example, an increase of 10°C for uricase (Chauhan et al., 2014) and 5°C for GOD (Figure 5.2; Kundu et al., 2012) has been reported. Immobilized ENPs show enhanced thermostability compared to that of free enzyme, for example, immobilized GOD NPs over free GOD. This could be due to the inter and intramolecular covalent cross-linkages which prevent conformational changes in the enzyme at higher temperature and thus enzyme deactivation (Kundu et al., 2012).

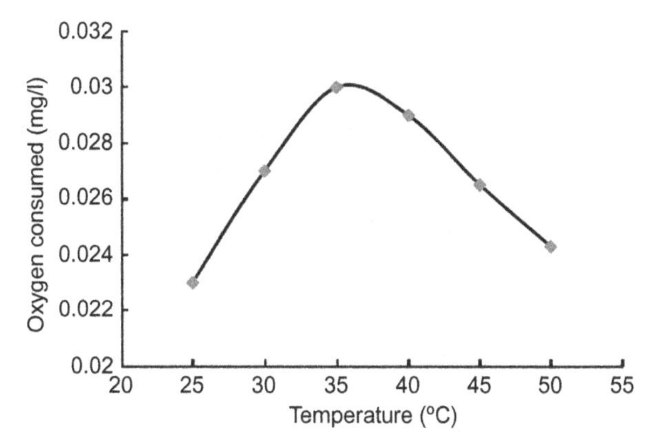

Figure 5.2 Effect of incubation temperature of reaction mixture on activity of glucose oxidase nanoparticles (GOD-NPs) immobilized onto nitro-cellulose membrane. The assay of immobilized GOD NPs was carried out as described in Figure 5.1 except that incubation temperature was varied as indicated in Figure 5.2. Source: Kundu et al., 2012.

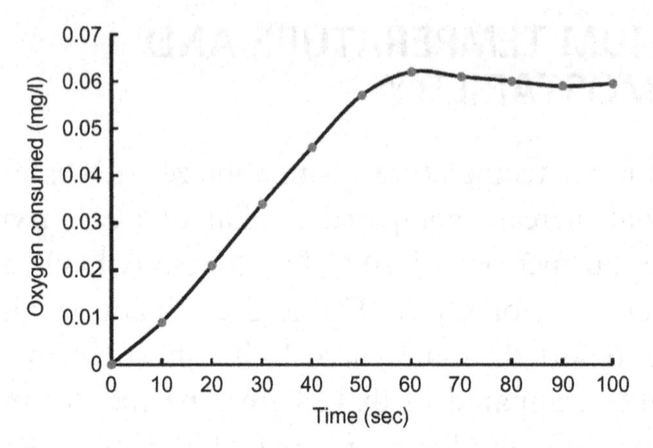

Figure 5.3 Effect of time of incubation on activity of glucose oxidase nanoparticles (GOD-NPs) immobilized onto nitro-cellulose membrane. The assay of immobilized GOD NPs was carried out as described in Figure 5.1 except that incubation time was varied as indicated in Figure 5.3. Source: Kundu et al., 2012.

RESPONSE TIME

The response time of immobilized ENPs decreases tremendously compared to that of free enzyme, for example, 17 times decrease for immobilized GOD NPs (Figure 5.3; Kundu et al., 2012), 39 times decrease for immobilized ChOx NPs (Chawla et al., 2013) and 43 times decrease for immobilized uricase NPs (Chauhan et al., 2014) than that of free enzyme.

Km VALUE

Apparent Km value of immobilized ENPs also decreases compared to that of free enzyme, for example, two times decrease in case of immobilized GOD NPs (Kundu et al., 2012), showing the higher affinity of ENPs towards their substrate after aggregation and immobilization, thus giving an advantage over free enzyme. However, apparent Km

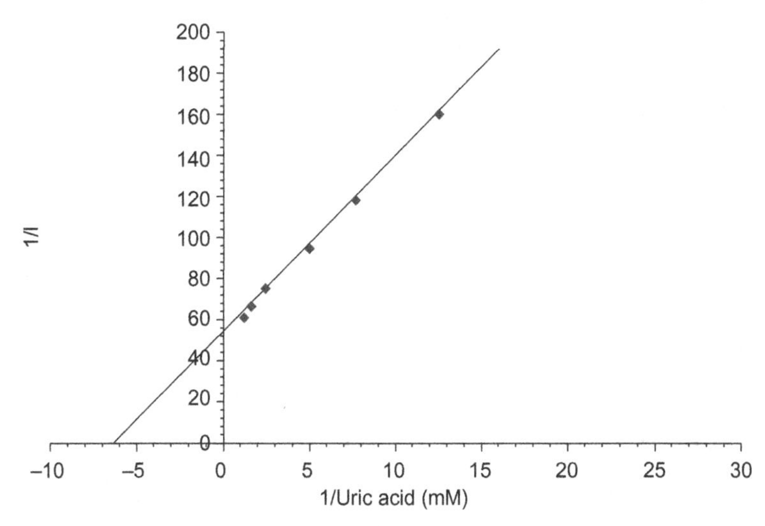

Figure 5.4 Lineweaver–Burke plot of the effect of uric acid concentration on response of uric acid biosensor based on uricase NPs modified Au electrode. Km was calculated using the following equation: $1/v = Km/Vmax \cdot 1/(S) + 1/Vmax$ *where* *v = Biosensor response in current (mAmp), S = Uric acid concentration in mM. Slope of LB plot = Km/Vmax; Intercept of the plot = 1/Vmax.*

value of uricase NPs bound to Au electrode as calculated from Lineweaver–Burke plot between reciprocal of uric acid concentration and reciprocal of biosensor response in current (mAmp) (Figure 5.4) was similar to that of free enzyme (Chauhan et al., 2014).

WORKING RANGE

The immobilized ENPs function in a quite lower working range compared to that of free enzyme (Table 5.1). Thus immobilized ENPs become more suitable for sensitive analytic purpose, than free enzyme.

STORAGE STABILITY

Immobilized ENPs show longer storage stability in cold, compared to that of free enzyme, revealing their superiority over free enzyme (Table 5.1).

REUSABILITY

Immobilized ENPs can be reused 100–200 times, which is not possible in case of free enzyme (Table 5.1). This economizes the use of enzymes in various fields.

SUMMARY

Immobilized ENP show higher optimum pH, incubation temperature, thermostability and storage stability in cold but lower response time, Km value and working range compared to those of free enzyme. Further, immobilized ENPs are reusable and thus more suitable and economic for various purposes, compared to free enzymes.

REFERENCES

Chauhan, N., Kumar, A., Pundir, C.S., 2014. Construction of an uricase nanoparticles modified Au electrode for amperometric determination of uric acid. App. Biochem. Biotechnol. Available from: http://dx.doi.org/10.1007/s12010-014-1097-6.

Chawla, S., Rawal, R., Sonia, Ramrati, Pundir, C.S., 2013. Preparation of cholesterol oxidase nanoparticles and their application in amperometric determination of cholesterol. J. Nanopart. Res. 15, 1934–1943.

Freitas, D.D.S., Spencer, P.J., Vasao, R.C., Abrahao-Neto, J., 2009. Biochemical and biopharmaceutical properties of PEGylated uricase. Int. J. Pharm. 387, 215–222.

Kalia, V., Goyal, L., Pundir, C.S., 1999. Determination of serum glucose with alkyl amine glass bound glucose oxidase. Chin. J. Biotechnol. 14, 201–204.

Kundu, N., Yadav, S., Pundir, C.S., 2012. Preparation and characterization of glucose oxidase nanoparticles and their application in DO metric determination of serum glucose. J. Nanosci. Nanotechnol. 13, 1710–1716.

Liu, G., Lin, Y., Ostatna, V., Wang, J., 2005. Enzyme nanoparticles based electronic biosensor. Chem. Commun. 27, 3481–3483.

Varma, R., Nene, S., 2003. Biosynthesis of cholesterol oxidase by *Streptomyces larendulae* NCIM 2421. Enzyme Microb. Technol. 33, 286–291.

Prado, D.D.S., [illegible], R., Vieira, R.C., Abrucio-Neto, [illegible], 2008. Biochemical and biopharmaceutical properties of PEGylated uricase. Int. J. Pharm. [illegible]

[illegible], V., [illegible], 1998. Determination of serum glucose with [illegible] glucose oxidase. [illegible]

[illegible]

[illegible]

[illegible]

CHAPTER 6

Applications of Enzyme Nanoparticles

Enzyme nanoparticles (ENPs) are used in the construction of improved biosensors. A biosensor is defined as an analytical device that incorporates a biological sensing element connected to a transducer to convert an observed response into a measurable signal, whose magnitude is directly proportional to the concentration of a specific chemical/biochemical or a set of chemicals in the samples (Figure 6.1). Immobilized oxido-reductases-based biosensors either consume oxygen, e.g. all the oxidases, or produce hydrogen peroxide (H_2O_2) (excluding oxidases that produce water) or generate reduced form of NAD(P)H, e.g. dehydrogenase, during the oxidation of the test sample. The former are dissolved oxygen (DO) metric, while the latter are amperometric biosensor. A biosensor has advantages over the other analytical devices in terms of their simplicity, rapidity, sensitivity and specificity. ENPs have been employed in the construction of DO metric/amperometric biosensors for detection of hydrogen peroxide, glucose, cholesterol and

Enzyme Nanoparticles. DOI: http://dx.doi.org/10.1016/B978-0-323-38913-6.00006-4

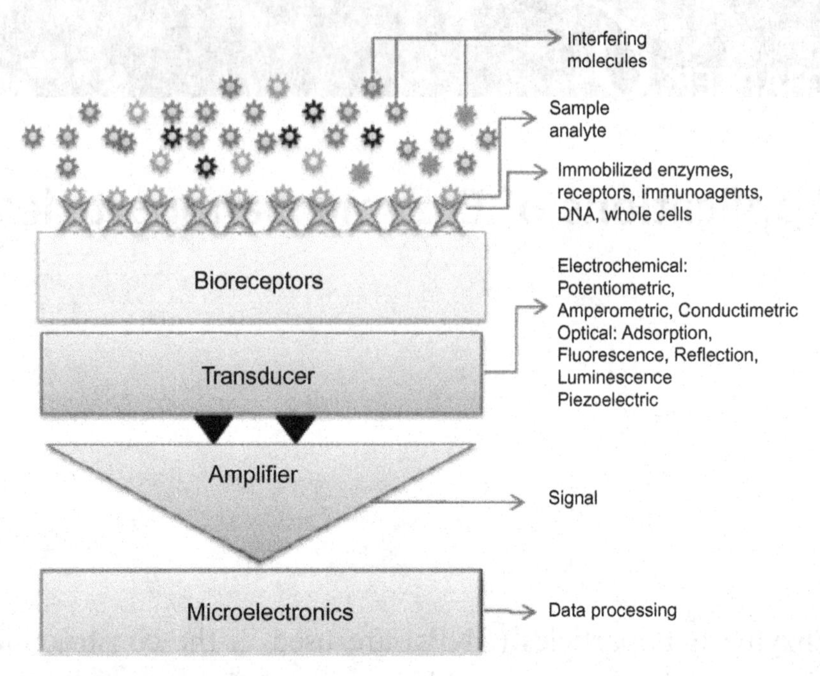

Figure 6.1 Concept of a biosensor.

uric acid in biological fluids (Liu et al., 2005; Sharma et al., 2011; Kundu et al., 2012; Chawla et al., 2013; Chauhan et al., 2014).

Direct adsorption of proteins/enzymes onto bulk metal surfaces frequently results in denaturation of protein and loss of bioactivity. In contrast, when the enzymes are adsorbed onto metal nanoparticles, and the enzyme-covered nanoparticles are then electrodeposited onto bulk electrode surface to construct biosensors, the bioactivity of enzymes is often retained. But when the ENPs (glucose oxidase, GOD NPs) are directly immobilized onto an electrode, it shows an improved analytical performance in terms of not just detection limits but also much larger current response (Sharma et al., 2011) (Figure 6.2). This chapter describes

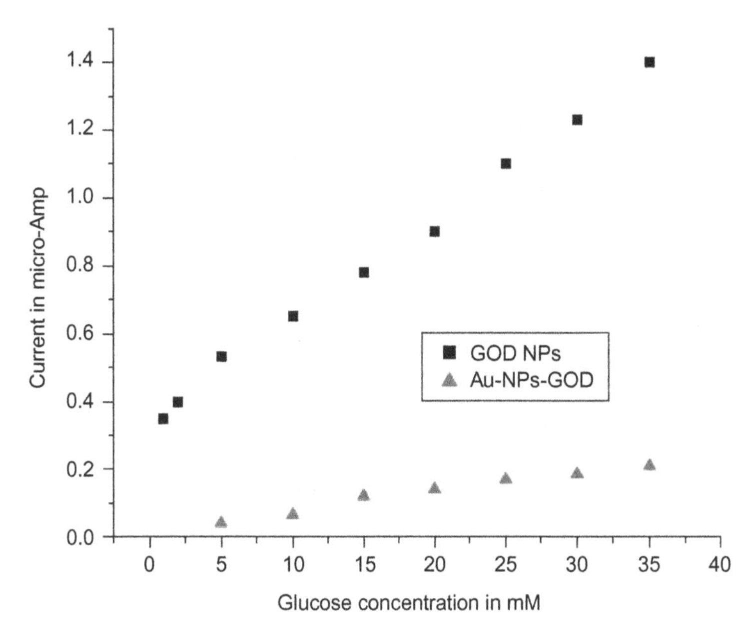

Figure 6.2 Variation of current as a function of glucose concentration using modified Pt electrode (▲) Au-NPs-GOD Complex and (■) GOD NPs. Source: Sharma et al. (2011).

the construction, application and evaluation of amperometric/DO metric biosensors based on immobilized ENPs of horseradish peroxidase (HRP), GOD, cholesterol oxidase (ChOx) and uricase and their comparison with earlier biosensors.

HORSERADISH PEROXIDASE (HRP) NPs-BASED BIOSENSOR

The determination of H_2O_2 is very important, as it is not only the product of highly selective oxidases but also employed in various fields such as food and pharmaceutical industries and environmental analysis. The higher concentration of H_2O_2 is also associated with diabetes,

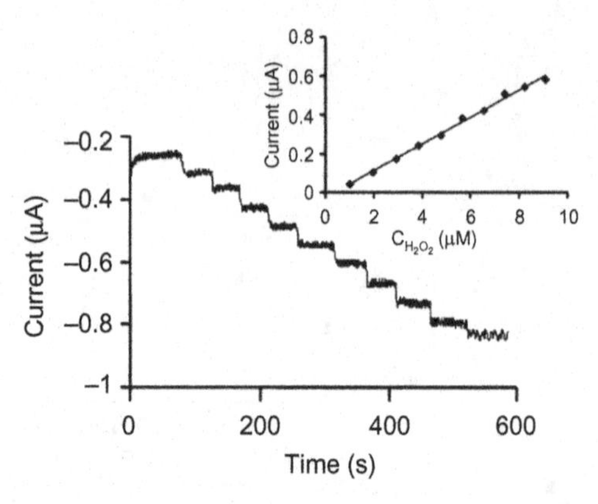

Figure 6.3 Typical steady-state response of the biosensor on successive injections of 0.1 mmol/l H_2O_2 into 1 ml of stirring PB. Applied potential, 2300 mV, supporting electrolyte, 0.1 mol/l pH 7.4 PB. Inset: the corresponding calibration plot. Source: Liu et al. (2005).

atherosclerosis and ageing, as it generates free hydroxyl radicals, which cause oxidative damage of the tissue components such as lipids and proteins, besides DNA (Lata et al., 2012). Liu et al. (2005) constructed for the first time an amperometric biosensor employing HRP NPs immobilized directly onto the surface of an Au electrode through thiol bond for detection of H_2O_2 without promoters and mediators. The electrochemical measurements were performed with a three-electrode system comprising a Pt wire as auxiliary electrode, Ag/AgCl as reference and HRP NPs/Au as working electrode. The biosensor showed optimum response within 5 s at 0.16 V and pH 7.4 in 0.1 M phosphate buffer. There was a linear relationship between biosensor response and H_2O_2 concentration in the range of 1–9 μM (Figure 6.3). The detection limit of the biosensor was 0.1 μM at a signal-to-noise ratio of 3. The coefficient of variation in detection of H_2O_2 concentration was 2.54% ($n = 10$).

GLUCOSE OXIDASE (GOD) NPs-BASED BIOSENSORS

Measurement of blood glucose levels is essential in the diagnosis and medical management of diabetes. Two types of glucose biosensors based on GOD NPs have been reported: amperometric (Sharma et al., 2011) and DO metric (Kundu et al., 2012). The former biosensor determines H_2O_2 generated from glucose by GOD NPs electrochemically (current in μAmp), while the latter measures the utilization of dissolved O_2 (in mg/l) in aerobic oxidation of glucose by GOD NPs. The following reactions occur in the measurement of their response:

$$\text{Glucose} + O_2 + H_2O \xrightarrow{\text{GOD}} \text{Gluconic acid} + H_2O_2$$

$$H_2O_2 \xrightarrow[\text{High voltage}]{} 2H^+ + O_2 + 2e^-(\text{Current})$$

Construction and Response Measurement

To construct an amperometric glucose biosensor, GOD NPs are immobilized covalently directly onto the amino-functionalized and glutaraldehyde activated Pt electrode to construct a GOD NPs/working electrode and then connecting it with Ag/AgCl as a reference electrode through micro-ammeter in a series at an applied potential of $0.4\ \text{V s}^{-1}$ across the two electrodes. In construction of DO metric glucose biosensor, GOD NPs are immobilized onto chitosan treated nitrocellulose (NC) membrane through glutaraldehyde coupling and then mounting this membrane

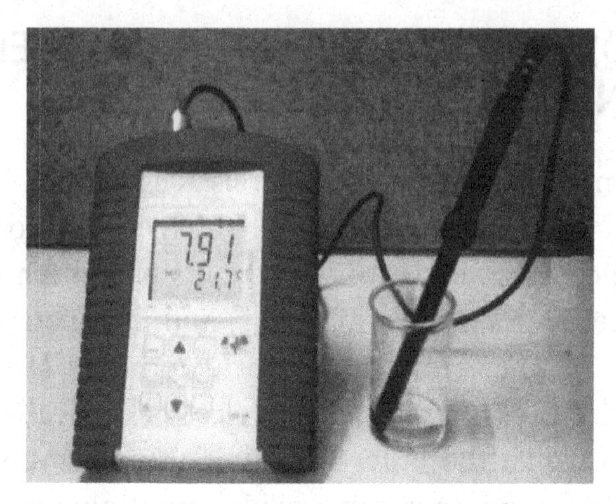

Figure 6.4 A dissolved oxygen (DO) meter with nitrocellulose membrane bound glucose oxidase.

over the sensing part of combined electrode of DO meter (Aqualytic OX53, Germany) with a parafilm (Figure 6.4). The three/combined electrode system is immersed into phosphate buffer 0.1 M, pH 7.4, in case of amperometric detection and 0.05 M, pH 6.0, saturated with O_2 bubbling for DO metric detection containing varying concentrations of glucose (10–500 mg/dl) and their responses (current in μA/ dissolved O_2 in mg/l) are measured (Figure 6.5). A comparison of the kinetic and analytic properties of both these biosensors is summarized in Table 6.1.

Biomedical Applications

The DO metric biosensor is employed for determination of serum glucose. The glucose content in serum samples as measured by this biosensor was in the range of 90–183 mg/dl in apparently healthy adults with a mean of 114.11 mg/dl and ranged from 200 to 258 mg/dl with a mean of 236.6 mg/dl in diabetic adults.

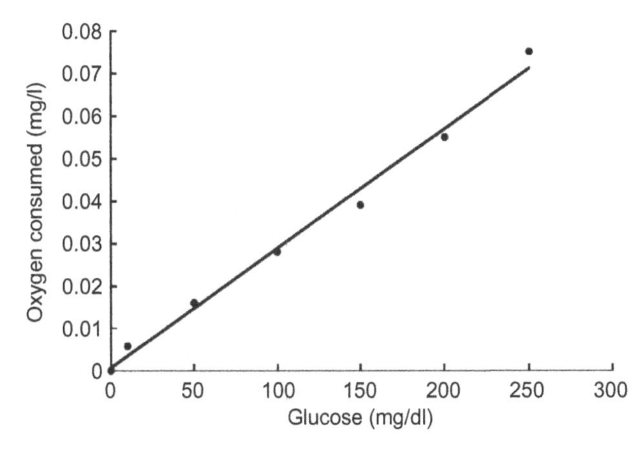

Figure 6.5 Calibration curve of glucose by DO metric glucose biosensor based on NC membrane bound GOD NPs aggregates. Source: Kundu et al. (2012).

Table 6.1 A Comparison of Analytical Parameters of Amperometric and DO Metric Glucose Biosensors Based on Glucose Oxidase Nanoparticles Immobilized onto Pt Electrode and NC Membrane

Properties	**Amperometric (Sharma et al., 2011)**	**DO Metric (Kundu et al., 2012)**
Optimum pH	7.4	6.0
Optimum temperature (°C)	27	30–45 (peak at 35)
Response time (s)	NR	60
Working range (Linearity) (mg/dl)	1–35	10–250
Apparent Km (mM)	NR	4.17
Vmax (mg/l)	NR	0.06
Limit of detection (mg/dl)	18	10
Reusability	NR	150
Storage stability at 4°C (days)	NR	180
Correlation with standard enzymatic colorimetric method by r value	NR	0.993

NR = Not reported

Evaluation

The working range of the DO metric biosensor is 10–250 mg/dl (0.56–1.39 mM), which is better than the earlier reported amperometric glucose biosensor based on colloid gold modified carbon paste electrode (0.04–0.28 mM) but comparable to the earlier reported glucose biosensors based on oxygen optrode (0.01–2 mM), electrocatalytically bulk modified epoxy graphite biocomposite (0.01–2 mM), oxygen principle electrodes (up to 1.83 mM), poly(neutral red) modified carbon film electrode (0.09–1.8 mM) and lower than based on poly[vinyl alcohol] matrix (0.05–3 mM), poly(*o*-phenylene diamine) layer (8–14 mM), Prussian blue layer (0.1–6.0 mM), chitosan-gold nanoparticle composite (0.005–2.4 mM), sol–gel-derived zirconium/nafion composite film (0.03–15.08 mM). The detection limit of biosensors is in the range of 1–10 mg/dl. There is a good correlation ($r = 0.993$) between serum glucose values of healthy and diabetic male adults under fasting conditions with those by standard enzymatic colorimetric method, using free GOD and peroxidase (Kundu et al., 2012) (Figure 6.6).

CHOLESTEROL OXIDASE (ChOx) NPs-BASED BIOSENSOR

Serum cholesterol determination is required for the diagnosis and medical management of hypercholesterolaemia, associated with pathological conditions like atherosclerosis, cardiovascular diseases, coronary artery diseases and transient ischaemic heart attacks. An amperometric cholesterol

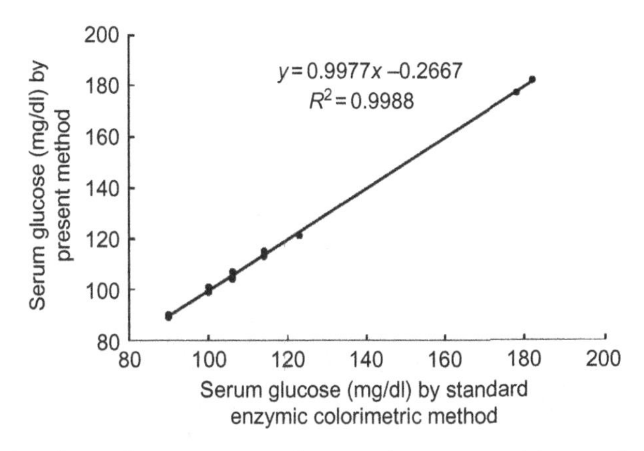

*Figure 6.6 Correlation between serum glucose values measured by standard enzymatic colorimetric method (*x-axis*) and the current method (*y-axis*) by DO metric glucose biosensor based on NC membrane bound GOD NPs aggregates.* Source: Kundu et al. (2012).

biosensor has been constructed based on ChOx NPs (Chawla et al., 2013). The following electrochemical reactions are involved in the measurement of response (current) of an amperometric cholesterol biosensor:

$$\text{Cholesterol} + O_2 + 2H_2O \xrightarrow{\text{ChOx NPs}} \text{Cholestene} + 2H_2O_2$$

$$2H_2O_2 \xrightarrow{0.27\ \text{V}} 4H^+ + 2O_2 + 4e^-$$

Construction and Response Measurement

To construct an amperometric cholesterol biosensor, thiolated ChOx NPs are immobilized onto Au wire and used as working electrode (ChOx NPs/Au electrode). This electrode along with Ag/AgCl as standard electrode and Pt wire as auxiliary electrode are connected through potentiostat/galvanostat (Autolab, Eco-Chemie, The Netherlands). The electrode is immersed into 0.1 M phosphate buffer (pH 7.0)

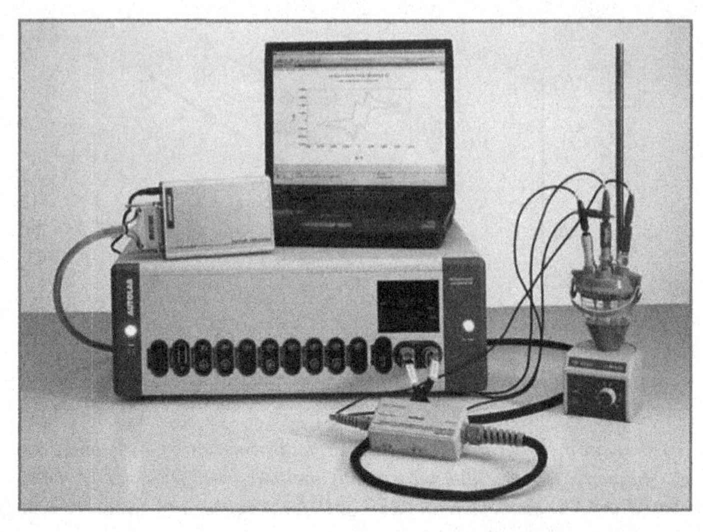

Figure 6.7 An amperometric cholesterol biosensor containing ChOx NPs modified Au electrode as working electrode, Ag/AgCl as reference and Pt wire as counter electrode connected through potentiostat/galvanostat.

containing varying concentrations of cholesterol and the response (current in μA) is measured in the potential range of 0.2–0.6 V s^{-1} (Figure 6.7).

The biosensor showed optimum response within 8 s at pH 6.0 and 35°C, when polarized at 0.27 V versus Ag/AgCl (Figure 6.8). The biosensor possesses high sensitivity and measures cholesterol concentrations as low as 1.56 mg/dl. The working/linear range was 12.5–700 mg/dl for cholesterol (Figure 6.7). The ChOx NPs/Au electrode lost 50% of its initial activity after its 180 uses during the span of 90 days, when stored dry at 4°C.

Biomedical Applications

The biosensor measured total cholesterol level in sera of apparently healthy adults and persons suffering from

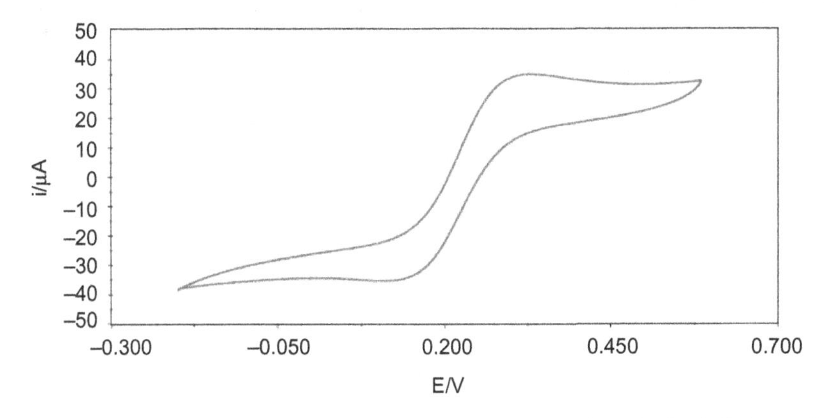

Figure 6.8 Cyclic voltammogram for ChOx NPs/Au electrode 100 μl of cholesterol in 0.1 M potassium PB (pH 7.0) with scan rate 20 mV/s. Source: Chawla et al. (2013).

various hypercholesterolaemic diseases and found in the range of 190–201 and 217.19–291.22 mg/dl, respectively (Table 6.2).

Evaluation

The biosensor showed a linear working range from 12.5 to 700 mg/dl for cholesterol (Figure 6.9), which is wider/better than earlier bionanosensors based on ChOx immobilized onto zinc oxide nanoparticles (2.0–11.0 mg/dl), poly(3)-TPAA.MnO_2 NPs/MWCNT (0.3–3.8 mg/dl) and graphene NPs (5.21–12.1 mg/dl). The detection limit (S/N = 3) was 1.56 mg/dl, which is also lower/better than these bionanosensors. The mean analytical recovery of cholesterol in serum (200 mg/dl) as measured by this sensor was 98%. The within and between batch CV for serum cholesterol determination were <2 and $<3\%$, respectively, revealing the high reproducibility and reliability of the biosensor. There was a good correlation (r = 0.99) between serum cholesterol values determined by standard enzymatic colorimetric

Table 6.2 Serum Cholesterol Levels of Apparently Healthy and Diseased Persons as Measured by Biosensor Employing ChOx NPs/Au as Working Electrode

Age Group ($n = 08$)	Sex		Serum Cholesterol (mg/dl) (Mean ± S.D.)	
	Healthy	Diseased	Healthy	Diseased
<10	M	M	153.16 ± 4.08	217.19 ± 9.37
	F	F	144.92 ± 8.06	202.91 ± 7.20
11–20	M	M	169.94 ± 7.78	233.44 ± 3.24
	F	F	164.82 ± 8.28	238.64 ± 7.01
21–30	M	M	188.18 ± 8.82	247.15 ± 5.95
	F	F	186.98 ± 8.04	241.20 ± 2.07
31–40	M	M	198.02 ± 3.28	264.99 ± 7.92
	F	F	189.82 ± 7.44	267.34 ± 6.11
41–50	M	M	208.56 ± 6.22	288.71 ± 1.37
	F	F	196.44 ± 8.18	297.93 ± 8.08
51–60	M	M	222.38 ± 8.46	283.45 ± 4.41
	F	F	218.78 ± 5.64	274.87 ± 9.92
61 & above	M	M	225.43 ± 8.82	291.22 ± 10.01
	F	F	225.12 ± 9.28	271.33 ± 11.97

The diseased persons were suffering from atherosclerosis, nephritis, diabetes mellitus, myxoedema and obstructive jaundice.
Source: Chawla et al. (2013).

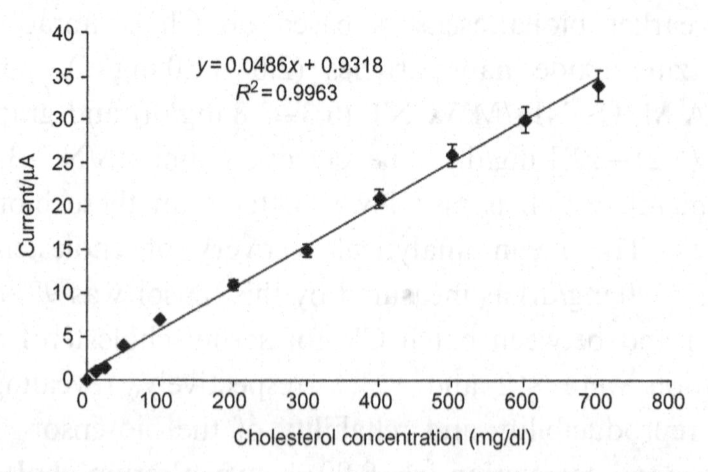

Figure 6.9 Effect of cholesterol concentration on the current response of the fabricated cholesterol biosensor based on ChOx NPs/Au electrode. Source: Chawla et al. (2013).

method and the present bionanosensor, showing the high accuracy of the biosensor. A number of metabolites such as glucose, fructose, ascorbic acid, uric acid, urea, acetone and bilirubin had negligible interference (Chawla et al., 2013).

URICASE NPs-BASED BIOSENSOR

Monitoring uric acid in blood or urine or both is very important, as it is a powerful indicator of early signs of kidney and metabolic disorders. Abnormal uric acid levels in the human body is caused by several diseases, e.g. gout, hyperuricaemia, Lesch-Nyhan syndrome, cardiovascular and chronic renal diseases. Among the various methods available for the measurement of uric acid, biosensing methods are comparatively more simple, rapid, specific, sensitive and user-friendly. An amperometric uric acid biosensor has been constructed employing uricase NPs (Chauhan et al., 2014). The following electrochemical reactions occur during the response (current) measurement of this biosensor:

$$\text{Uric acid} + O_2 + H_2O \xrightarrow{\text{Uricase NPs}} \text{Allanoin} + H_2O_2$$

$$H_2O_2 \xrightarrow{0.2\ \text{V}} 2H^+ + O_2 + 2e^-$$

Construction and Response Measurement

To construct an amperometric uric acid biosensor, thiolated uricase NPs are immobilized covalently on the surface of Au wire and used as uricase NPs electrode/working electrode. This electrode along with Ag/AgCl as standard electrode and Pt wire as auxiliary electrode are connected through potentiostat/galvanostat. The three-electrode system is immersed into 0.05 M Tris HCl buffer (pH 8.5) containing varying concentrations of uric acid (0.1–1.0 mM) and the response (current in μAmp) is measured in the potential range of 0–1.5 V s^{-1}. The biosensor showed maximum response at 0.2 V, within 7 s at pH 8.5 and 40°C, and detected uric acid level as low as 5.0 μM at a signal-to-noise ratio of 3. The biosensor had a linear working range, 0.005 to 0.8 mM for uric acid with a sensitivity of 0.03 mA μM^{-1} cm^{-2}. Apparent Km value for uric acid was 0.058. The biosensor retained 85% of its initial activity over a period of 7 months, when stored dry at 4°C.

Biomedical Applications

The biosensor measured uric acid levels in serum and urine in apparently healthy adults and diseased persons. Serum uric acid was in the range of 2.6 to 5.73 mg/dl with a mean of 4.58 mg/dl in healthy adults and ranged from 6.9 to 11.9 mg/dl with a mean of 8.803 mg/dl ($n = 10$) in diseased persons suffering from gout, leukaemia, toxaemia of pregnancy and nephrolithiasis (Table 6.3). The urine uric acid levels in healthy adults ranged from 2.7 to 5.4 mg/dl with a mean of 3.68 mg/dl and in diseased persons from 7.42 to 12.91 mg/dl with a mean of 10.20 mg/dl (Table 6.4).

Evaluation

The biosensor showed a linear relationship between current (mA) and uric acid concentration ranging from 0.005 to

Table 6.3 Serum Uric Acid Levels in Apparently Healthy Individuals and Diseased Persons as Measured by Biosensor Based on Uricase NPs Modified Au Electrode

S. No.	Sex	Serum Uric Acid (mg/dl) in Apparently Healthy Individuals	Serum Uric Acid (mg/dl) in Diseased Persons
I	M	2.6	7.28
2	M	3.8	6.9
3	F	4.8	9.58
4	F	6.7	8.62
5	F	5.7	7.4
6	M	3.1	8.9
7	F	5.4	10.0
8	M	4.3	11.9
9	M	3.7	8.5
10	F	5.73	8.95
		Mean = 4.58	Mean = 8.803

Source: Chauhan et al. (2014).

Table 6.4 Urine Uric Acid Levels in Apparently Healthy Individuals and Diseased Persons as Measured by Biosensor Based on Uricase NPs Modified Au Electrode

S. No.	Sex	Urine Uric Acid (mg/dl) in Apparently Healthy Individuals	Urine Uric Acid (mg/dl) in Diseased Persons
1	M	3.6	14.22
2	M	3.8	8.91
3	F	3.8	8.58
4	F	2.7	11.62
5	F	3.7	7.42
6	M	3.1	8.90
7	M	5.4	11.00
8	M	4.3	12.91
9	M	3.7	9.50
10	F	2.73	8.95
		Mean = 3.68	Mean = 10.20

Source: Chauhan et al. (2014).

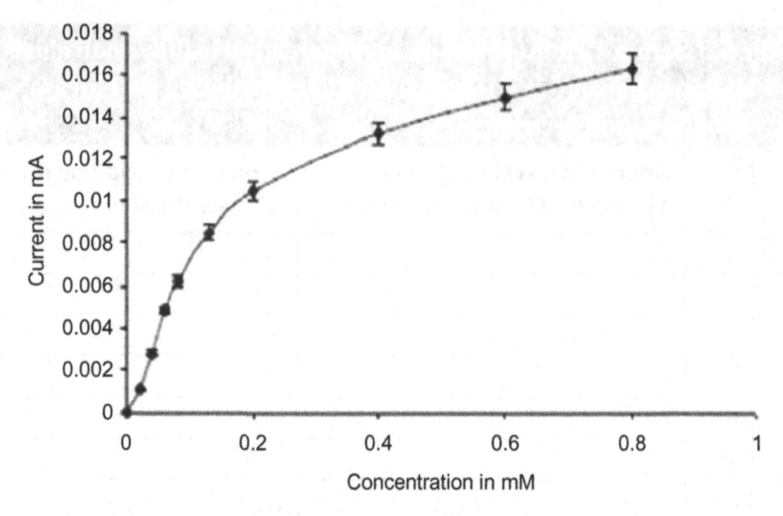

Figure 6.10 Effect of uric acid concentration on response of uric acid biosensor based on uricase NPs modified Au electrode. Source: Chauhan et al. (2014).

Table 6.5 Analytical Recovery of Added Uric Acid in Serum as Measured by Amperometric Biosensor Based on Uricase NPs/Au Electrode

Uric Acid Added (mg/l)	Uric Acid Found	% Recovery
Nil	37.0	–
10	46.8	99.57
20	56.3	98.77

Source: Chauhan et al. (2014).

0.8 mM with a sensitivity of 0.03 mA μM^{-1} cm^{-2} in Tris HCl buffer (pH 8.5) (Figure 6.10), which is better than the earlier reported biosensors. The minimum detection limit of the biosensor was 5.0 μM at a signal-to-noise ratio of 3, which is also better/lower than any other earlier biosensors. Analytical recovery of exogenously added uric acid in serum (10 and 20 mg/l) in the method was 99.57% and 98.77%, respectively, showing the liability of the method (Table 6.5). Uric acid content in the same serum sample

Table 6.6 Precision for Determination of Uric Acid in Sera by Amperometric Biosensor Based on Uricase NPs/Au Electrode

N (Number of Sample)	Uric Acid (mmol/l) (Mean ± S.D.)	Coefficient of Variation (CV) (%)
Within assay (5)	69.3 ± 3.90	5.6
Between assay (5)	58.0 ± 2.78	4.7

Source: Chauhan et al. (2014).

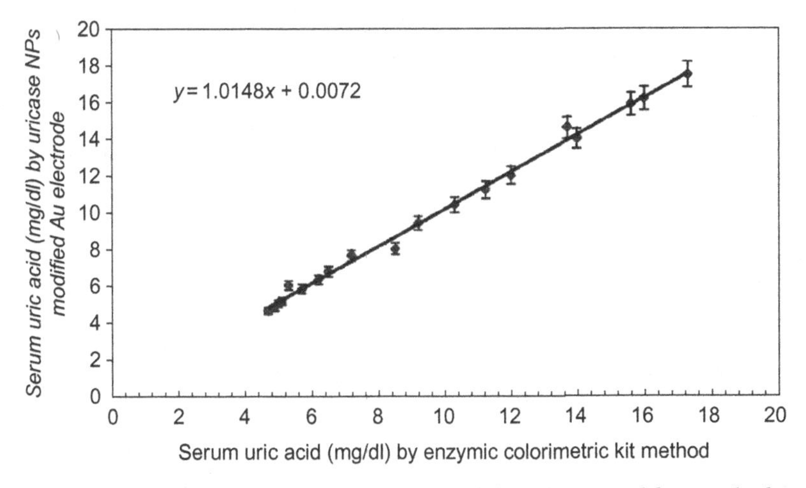

Figure 6.11 Correlation between serum uric acid values determined by standard enzymatic colorimetric method (x) *and the present uricase NPs modified Au electrode* (y). Source: Chauhan et al. (2014).

was determined five times on a single day (within batch) and again after storage at −20°C for 1 week (between batch). Within and between batch coefficients of variation (CV) for serum uric acid determinations were <5.6% and <4.7%, showing the high repeatability and reproducibility of the method (Table 6.6). The serum uric acid values as measured by the biosensor showed a good correlation (r = 0.998) with those by standard enzymatic colorimetric method (Figure 6.11), revealing the high accuracy of the method. The biosensor was unaffected by physiological concentrations of glucose, cholesterol, ascorbic acid, urea, pyruvate, bilirubin, NADH, FMN, riboflavin and $MnCl_2$ (Chauhan et al., 2014).

SUMMARY

ENPs are employed in the construction of improved amperometric biosensors for various metabolites. These ENPs-based biosensors measured H_2O_2, glucose, cholesterol and uric acid in biological fluids, which is required in the diagnosis and medical management of different diseases.

REFERENCES

Chauhan, N., Kumar, A., Pundir, C.S., 2014. Construction of an uricase nanoparticles modified Au electrode for amperometric determination of uric acid. App. Biochem. Biotechnol. 174: 1683–1694.

Chawla, S., Rawal, R., Sonia, Ramrati, Pundir, C.S., 2013. Preparation of cholesterol oxidase nanoparticles and their application in amperometric determination of cholesterol. J. Nanopart. Res. 15, 1934–1943.

Kundu, N., Yadav, S., Pundir, C.S., 2012. Preparation and characterization of glucose oxidase nanoparticles and their application in DO metric determination of serum glucose. J. Nanosci. Nanotechnol. 13, 1710–1716.

Lata, S., Batra, B., Karwasra, N., Pundir, C.S., 2012. An amperometric biosensor basd on cytochrome C immobilizedonto nickel oxide nanoparticles/carboxylated multiwalled carbon nanotubes/polyaniline modified gold electrode. Process Biochem. 47, 992–998.

Liu, G., Lin, Y., Ostatna, V., Wang, J., 2005. Enzyme nanoparticles based electronic biosensor. Chem. Commun. 27, 3481–3483.

Sharma, S., Shrivastav, A., Gupta, N., Srivastava, S., 2011. Amperometric biosensor: Increased sensitivity using enzyme nanoparticles. In: 2010 International Conference on Nanotechnology and Biosensors, IPCBEE. vol. 2, pp. 24–26.

CHAPTER 7

Future Development in Enzyme Nanoparticles

The enzymes (both in free as well as immobilized form) have been employed in various fields of our day to day life, for example, cellulases, β-glucanases, α-glucanases, proteases, maltogenic amylases, renin (protease), lactase, protease, catalases, pectinase, amyloglucosidase, protease, papain-α-amylases, amyloglycosidases, protease, maltogen amylase (novamyl), pentosanase, glucose oxidase, protease, trypsin, aminopeptidases, α-amylase, glucoamylase, hemicellulases, maltogenic amylases, glucose isomerases, inulinases, β-glucanases, xylanses, phytase, chymosin and pepsin in food industries; lipases, proteases, amylases and cellulases in detergent industries; proteases, cellulases, catalases, laccase, xylanase and phytase in textile industries; lipase in leather industry, xylanases and cellulases in paper and pulp industry; proteinase, lipase, glucoamylase and glucose oxidase in personal care product; cellulases as biofuels, ligase, DNA polymerase and nuclease (restriction

Enzyme Nanoparticles. DOI: http://dx.doi.org/10.1016/B978-0-323-38913-6.00007-6

enzymes) in recombinant DNA technology; asparginase, collagenase, glutaminase, hyaluronidase, lysozyme, rhodanase, ribonuclease, β-lactamase, streptokinase, trypsin, uricase; urokinase in treatment of various disorders and protease in photography; alcohol dehydrogenase, bilirubin oxidase and peroxidase, glucose oxidase and peroxidase, lactate oxidase and peroxidase, oxalate oxidase and peroxidase, uricase and peroxidase, cholesteryl esterase, cholesterol oxidase and peroxidase, lipase, glycerol kinase, glycerol-3-phosphate oxidase and peroxidase, 3-α-hydroxysteroid dehydrogenase and diaphorase, creatininase, creatinase, sarcocine oxidase and peroxidase in enzo kits for determination of various metabolites in biological materials required for diagnosis and medical management of different diseases (Appendix). A large number of oxido reductases and hydrolases in their native form have been employed for construction of different enzyme sensors required for simple, fast, highly sensitive detection of various important substances.

Liu et al. (2005) for the first time prepared the nanoparticles of horseradish peroxidase by desolvation with ethanol, cross linked with glutaraldehyde, functionalized with cysteine and immobilized them directly onto Au electrode for construction of a reagentless amperometric H_2O_2 biosensor without promoters and mediators. Sharma et al. (2011) prepared and immobilized GOD NPs on Pt electrode to construct amperometric glucose biosensor, which showed increased activity and longer stability compared to enzyme bound to nanoparticles. Thereafter, the nanoparticles of few more enzymes such as cholesterol oxidase (ChOx) (Chawla et al., 2013) and uricase (Chauhan et al.,

2014) were prepared using the same method and employed in construction of improved amperometric biosensors for determination of glucose, cholesterol and uric acid in biological fluids. These biosensors showed an improved analytic performance not only in terms of detection limit but also larger current response and longer shelf life. This new immobilization method of enzyme, that is, the direct immobilization of enzyme nanoparticles with proper choice of functionalization seems to be a promising strategy for designing biosensors/biomedical devices, biofuel cells and enzyme reactors. The functionalized ENPs could also be employed as novel tags for bioaffinity assays of protein and DNA. This new biosensor technology is thus expected to open new opportunities for biosensors, clinical diagnostics and medical management. The ENPs based sensor could also be miniaturized within mini biochips to provide small, portable, user friendly, low cost and highly sensitive diagnostic devices. The increased stability of ENPs could greatly expand enzymatic catalysis at elevated temperature. This could be used to increase the shelf life of biosensors. Although nanoparticles of proteins (bovine serum albumin) have been prepared by emulsification in plant oil, desolvation by ethanol or natural salts followed by glutaraldehyde cross-linking, simple coacervation by anhydrous ethanol, glutaraldehyde and ethanol amine and cross-linking in water and oil emulsion under high pressure, the ENPs were prepared by desolvation method only (Sailaja et al., 2011). However, other methods could also be employed to prepare the ENPs.

Iranian scientist (Afzal Karimi, member of the scientific board of university of Tabriz, posted on August, 2, 2011, from Tehran) used GOD NPs to deoxygenate water fed to steam boilers. This was a better method than chemical method using hydrazine for oxygen removal from water, as it was environmentally friendly, safe and economic. Thus ENPs could also be used for water purification.

It is expected the use of ENPs in place of free enzymes/immobilized enzymes in the various industries as mentioned above would provide the better results. This field is opening its arms from the last decade and little work has been done in this area. So, there is a wide scope for research in the area of enzyme nanoparticle technology.

REFERENCES

Chauhan, N., Kumar, A., Pundir, C.S., 2014. Construction of an uricase nanoparticles modified Au electrode for amperometric determination of uric acid. App. Biochem. Biotechnol. Available from: http://dx.doi.org/10.1007/s12010-014-1097-6.

Chawla, S., Rawal, R., Sonia, Ramrati, Pundir, C.S., 2013. Preparation of cholesterol oxidase nanoparticles and their application in amperometric determination of cholesterol. J. Nanopart. Res. 15, 1934–1943.

Liu, G., Lin, Y., Ostatna, V., Wang, J., 2005. Enzyme nanoparticles based electronic biosensor. Chem. Commun. 27, 3481–3483.

Sailaja, A.K., Amareshwar, P., Chakarvarty, P., 2011. Different techniques used for preparation of nanoparticles using natural polymers and their application. Int. J. Pharm. Pharm. Sci. 3, 45–50.

Sharma, S., Shrivastav, A., Gupta, N., Srivastava, S., 2011. Amperometric biosensor: increased sensitivity using enzyme nanoparticles. In: 2010 International Conference on Nanotechnology and Biosensors, IPCBEE, vol. 2, pp. 24–26.

APPENDIX

Enzymes Used in Day-to-Day Life

Table 1 Enzymes Used in Food Industries

Industry	Enzyme(s)	Function(s)
Brewing	Cellulases	For liquefaction, clarification and supplementation of malt enzymes
	β-Glucanases	
	α-Glucanases	
	Proteases	
	Maltogenic amylases	
Dairy	Renin (protease)	Coagulant in cheese production
	Lactase	Hydrolysis of lactose
	Protease	Hydrolysis of whey protein
	Catalases	Removal of H_2O_2
Wine and fruit juices	Pectinase	Increase of yield and juice clarification
Alcohol production	Amyloglucosidase	Conversion of starch to sugar
Meat	Protease	Meat tendering
	Papain	
Baking	α-amylases	Breakdown of starch
	Amyloglycosidases	Saccharification
	Protease	Breakdown of proteins
	Maltogen amylase (Novamyl)	Delay process by which bread becomes stale
	Pentosanase	Breakdown of pentosan, leading to reduced gluten production
	Glucose oxidase	Stability of dough
Protein	Protease	Breakdown of various components
	Trypsin	
	Aminopeptidases	

(*Continued*)

Table 1 (Continued)

Industry	Enzyme(s)	Function(s)
Starch	α-Amylase	Modification and conversion of starch
	Glucoamylase	
	Hemicellulases	
	Maltogenic amylases	
	Glucose isomerases	Fructose formation
Inulin	Inulinases	Production of fructose syrups
Animal feed	β-Glucanases	Added in barley-based feed diet
	Xylanases	Fibre solubility
	Phytase	Release of phosphate
Cheese making	Chymosin	Milk warming
	Pepsin	

Table 2 Enzymes Used in Detergent Industry

Enzyme(s)	Functions
Lipase	Removal of fatty stains by decomposing into water soluble compounds
Proteases	Removal of protein contamination
Amylases	Removal of starch-based stains
Cellulases	Colour brightening and softening

Table 3 Enzymes Used in Textile Industry

Enzyme(s)	Function(s)
Proteases	Dehairing of animal hides by minimizing the physical damage to fibres/biostaining
Cellulases	De-greasing and fabric brightening (biopolishing)
Catalases	Treating cotton fibres to prepare them for dyeing process
Laccase	Breaking down of dirt to clean the clothes more effectively
Xylanase	Solubilization of fibres
Phytase	Releasing of phosphate

Table 4 Enzyme Used in Leather Industry

Enzyme	Function(s)
Lipase	Batting phase (degradation of collagen to make leather soft and easy to dye)

Table 5 Enzyme Used in Paper and Pulp Industry

Enzyme(s)	Function(s)
Xylanases	Pulp bleaching by hydrolyzing residual xylan to liberate lignin
	Removal of fine particles from pulp
Cellulases	Pre-treatment of wood and breakdown of lignin fibres

Table 6 Enzymes Used in Personal Care Products

Enzyme(s)	Function(s)
Proteinase	Cleaning of contact lenses
Lipase	
Gluco-amylase	Disinfectants in toothpaste
Glucose oxidase	

Table 7 Enzyme Used in Biofuels

Enzyme	Function
Cellulases	Conversion of cellulose fibrils into sugar for its conversion into ethanol by micro-organisms

Table 8 Enzymes Used in Recombinant DNA Technology

Enzymes	Function
Ligase	Joining of DNA fragments
DNA polymerase	Joining of nucleotides during DNA replication
Nuclease (restriction enzymes)	Cleavage of nucleic acid at some specific site

Table 9 Therapeutic Uses of Enzymes in Treatment of Diseases/Disorders

Enzyme	Disease/Disorder
Asparginase	Leukaemia
Collagenase	Skin ulcers
Glutaminase	Leukaemia
Hyaluronidase	Heart attacks
Lysozyme	Antibiotic
Rhodanase	Cyanide poisoning
Ribonuclease	Antiviral
β-Lactamase	Penicillin allergy
Streptokinase	Blood clots
Trypsin	Inflammation
Uricase	Gout
Urokinase	Blood clots

Table 10 Enzyme Used in Photography

Enzyme	Function
Protease	Dissolving gelatin off scrapfilm, allowing recovery of its silver content

Table 11 Enzymes Used in Pharmaceutical Products – Enzo Kits

Enzymes	Enzo Kits
Urease	Urea
3-α-Hydroxysteroid dehydrogenase and diaphorase	Bile acids
Bilirubin oxidase and peroxidase	Bilirubin
Glucose oxidase and peroxidase	Glucose
Lactate oxidase and peroxidase	Lactic acid
Oxalate oxidase and peroxidase	Oxalic acid
Uricase and peroxidase	Uric acid
Cholesteryl esterase, cholesterol oxidase and peroxidase	Cholesterol
Glycerol kinase, glycerol-3-phosphate oxidase and peroxidase	Glycerol
Lipase, glycerol kinase, glycerol-3-phosphate oxidase and peroxidase	Triglyceride
Creatininase, creatinase, sarcocine oxidase and peroxidase	Creatinine

STATISTICAL FORMULAE FOR PRECISION AND CORRELATION

To evaluate the data obtained by present methods following statistical formulae were used:

Standard deviation (σ)

$$(\sigma) = \sqrt{\frac{\Sigma(X-\overline{X})^2}{(n-1)}}$$

where

X = each score; $\overline{X}$ = mean; n = number of samples

Coefficient of variation

$$\text{Coefficient of variation} = \frac{\sigma \times 100}{\alpha}$$

where

σ = SD; α = means of series

Correlation coefficient (r)

$$r = \frac{n\Sigma xy - \Sigma x \Sigma y}{\{n\Sigma x^2 - (\Sigma x)^2\}\{n\Sigma y^2 - (\Sigma y)^2\}}$$

where

x = value obtained by reference method, y = values obtained by present method

Printed in the United States
By Bookmasters